Lecture Notes in Statistics

Edited by J. Berger, S. Fienberg, J. Gani, K. Krickeberg, I. Olkin, and B. Singer

70

Kōji Iida

Studies on the Optimal Search Plan

Springer-Verlag
Berlin Heidelberg New York London Paris
Tokyo Hong Kong Barcelona Budapest

Author
Kōji Iida
Department of Applied Physics
The National Defense Academy
Yokosuka 239, Japan

Mathematical Subject Classification: 90B40

ISBN-13: 978-0-387-97739-3 e-ISBN-13: 978-1-4612-2876-9
DOI: 10.1007/978-1-4612-2876-9

Typesetting: Camera ready by author

47/3140-543210 – Printed on acid-free paper

Preface

This monograph brings together the author's studies on the problem of optimal allocation of searching effort to detect a target. Except for the final chapter, the allocation problems of searching effort dealt with are one-sided, in the sense that the action of the target is determined stochastically in advance and the searcher chooses how to allocate his available searching effort. In these problems, the searcher is assumed to have some knowledge about the target's location in the form of a probability distribution, called the target distribution. The searcher's sensor capability of detecting the target, the detection law, is a function relating the probability of detecting the target to the searching effort applied in the target position and is also assumed to be given. The primary focus of this monograph is on problems in which the searching effort is arbitrarily divisible in allocating it to the target space.

This monograph is composed of five chapters with an introductory chapter. In the introductory chapter, we begin by providing a short survey of search theory and give a sketch for the various attributes of search models in modeling search operations. A chapter-by-chapter summary of this monograph is also given. Chapter 1 deals with the optimal search plan, which minimizes the expected risk of the search for a stationary target. The optimal search problems for a moving target are studied in Chapter 2. In Chapter 3, we investigate the optimal whereabouts search plans minimizing the expected risk. In this chapter, we deal with the whereabouts search problems for two types of targets, mortal and immortal. The optimal search and stop problem in two-stage search is investigated in Chapter 4. In the final chapter, we analyze the forestalling detection probability in two-sided search-and-search problems in a broad search and a barrier search. In each chapter, a short survey of related studies presented before is given in the first section. The references cited are at the end of the monograph.

The author's studies presented in this monograph were reported until the beginning of 1990 in the following journals: *Journal of the Operations Research Society of Japan*, *Naval Research Logistics*, *Mathematica Japonica*, *Memoirs of the National Defense Academy*, *Keiei Kagaku*, *Defense Technology Journal*, and *Scientific and Engineering Reports of the National Defense Academy*. (The reports in the last three journals are written in Japanese.) The original version of this monograph was submitted in partial fulfillment of the requirement for the degree of Doctor of Engineering from Ōsaka University, December 1988. This monograph also contains several numerical examples and discussions in addition to the original one.

Kōji Iida
June 1990

Acknowledgments

This work was done under the guidance of Professor Takasi Kisi of the National Defense Academy and Professor Minoru Sakaguchi of Ōsaka University.

The author would like to express his deep gratitude to Professor Takasi Kisi of the National Defense Academy, who introduced him to the study of search theory and has given him continuous encouragement.

The author would like to express his sincere appreciation to Professor Minoru Sakaguchi of Ōsaka University for supervising this work. Without Professor Sakaguchi's invaluable criticisms and warm encouragement, the author would have been unable to accomplish the work.

The author is also indebted to Dr. Teruhisa Nakai of Ōsaka University, who gave him a number of precious suggestions.

Contents

Preface

Acknowledgments

Introduction

Short Survey of Search Theory 1

Attributes of Search Models 2

Outline of Chapters 5

Chapter 1. Optimal Search Plan for a Stationary Target Minimizing the Expected Risk

1.1. Introduction 8

1.2. Optimal Search Plan

1.2.1. Optimal Distribution of Searching Effort 10

1.2.2. Optimal Stopping Time of the Search 18

1.3. Two-Box Problem 22

Chapter 2. Optimal Search Plan for a Moving Target

2.1. Introduction 29

2.2. Optimal Search Plan Maximizing the Detection Probability 30

2.3. Optimal Search Plan Minimizing the Expected Risk

2.3.1. Optimal Distribution of Searching Effort 33

2.3.2. Optimal Stopping Time of the Search 40

2.3.3. A Numerical Example 43

2.4. Optimal Search Plan for a Target with Conditionally Deterministic Motion 45

Chapter 3. Optimal Whereabouts Search Plan Minimizing the Expected Risk

3.1. Introduction 60

3.2. Optimal Whereabouts Search Plan for an Appeared Target

3.2.1. Optimal Distribution of Searching Effort 61

3.2.2. Optimal Stopping Time of the Search 66

3.2.3. Optimal Box to Be Guessed 69

3.3. Optimal Whereabouts Search Plan for a Mortal Target 70
3.4. Numerical Examples 80

Chapter 4. Optimal Investigating Search Plan for Contacts in Two-Stage Search
4.1. Introduction 84
4.2. Optimal Investigating Search Plan
4.2.1. Optimal Selection of Contacts for Investigation and Optimal Stopping Time 86
4.2.2. Properties of the Optimal Investigating Search Plan 94
4.3. Numerical Examples 96

Chapter 5. Forestalling Detection in Two-Sided Search
5.1. Introduction 103
5.2. Forestalling Detection in Broad Search 104
5.3. Forestalling Detection in Barrier Search 115

Bibliography 122

Author Index 126

Subject Index 128

Introduction

Search theory, one of the oldest areas of operations research, has continuously provided powerful support for planning efficient search operations in real-world applications. The aim of the searcher is to efficiently find the target in his search operations, and analysts are asked to offer theoretical bases for optimal planning of the search. The search theory makes many valuable contributions to the study of the optimal solutions of search problems by applying optimization techniques, such as nonlinear programming, fractional programming, dynamic programming, calculus of variations, and the maximum principle of optimal control theory. We begin by providing a short survey of search theory.

Short Survey of Search Theory

During World War II, B.O. Koopman and his colleagues in the Operations Research Group of the U.S. Navy devoted intensive efforts to the development of efficient search operations in antisubmarine warfare. They established many of the fundamental concepts of search theory such as lateral range curve, sweep width, sweep rate, coverage factor, random search, and parallel sweep. They also obtained basic theorems for the optimal distribution of searching effort, maximizing the probability of detection of a stationary target. Their studies formed the subject of "Search and Screening" by Koopman (1946) and were reported partially in the journal of the Operations Research Society of America (Koopman 1956a, 1956b, 1957). An updated and expanded version of the book was published later by Koopman (1980). After the war, the study of search theory made rapid progress and grew to become a major field within operations research. The theoretical development of the optimal distribution of searching effort until the first half of the 1970s was summarized by Stone (1975).

The principles of search theory can be applied to many important search operations. Possible areas of application include the search for and rescue of a lost party in a mountain or a missing boat on the ocean, the surveillance of frontiers or territorial seas, the search for mineral deposits, medical diagnosis, and the search for malfunctions in an industrial process. New problems in and potential applications of search theory were discussed at the NATO Advanced Research Institute on Search Theory and Applications held in Praia Da Rocha, Portugal, in 1979. The proceedings of this meeting, which contained the presented papers and workshop reports were edited by Haley and Stone and published in 1980. Reviews of the current status of search theory were given by Stone (1983) and Richardson (1987). According to these reports, search theory still remains a field of active research in spite of the considerable advances made since its early study. Many problems remain to be solved, particularly in cases involving false targets and multiple true targets. More detailed surveys on search theory and bibliographies can be found in Enslow (1966), Dobbie (1968), Moore (1970), Kisi and Iida (1971), Iida (1974), Stone (1983), and Richardson (1987). The main books on search theory are Koopman (1946,1980), Tada (1973), Stone (1975), Gal (1980), Haley and Stone (Eds. 1980), Washburn (1981a), Ahlswede and Wegener (1987), and Chudnovsky,D. and Chudnovsky,G. (Eds. 1989).

Attributes of Search Models

Theoretical studies in search theory can be classified according to the assumptions made about the target space, target motion, divisibility of the searching effort, detection law, number of decision time points, and measures of effectiveness. A set of those assumptions specifies the search model.

Studies of search theory can be classified broadly into two parts; one-sided search and two-sided search. In the former search model, the action of the target is determined stochastically in advance; in the latter, the target is allowed to take action against the searcher. If the searcher can decide on his action at several time points, on the basis of the information obtained in the previous search steps, the search process is called a sequential search.

The space in which both the searcher and the target are operating is called the target space. In some cases, the target space consists of discrete points, called boxes; in others, the space is a continuum. If the target appeared in the target space before the beginning of the search and stays there during the search, it is called the appeared target. On the other hand, if the target appears or disappears during the search, it is called either an appearing or disappearing target. If the target is an appeared target and the probability distribution of the target position in the target space does not vary during the search, the target is called a stationary target. If the target moves in the target space, it is called a moving target.

The searcher's available resources such as man-hours or expenditure for the search, are called searching effort. If the searching effort is quantized, it is called discrete searching effort; if it is continuously divisible, it is called continuous searching effort. In problems involving discrete searching effort, the target space is usually considered a discrete space. In the case of discrete effort, allocation of the unit searching effort in any box is called a look with a conditional detection probability α_i; the target is detected with probability α_i by the look, given that the target is in box i. In the case of continuous effort, we define a detection function $b(x,z)$, which gives the conditional probability of detecting the target with density z of searching effort applied in a position x given that the target is really at x. The function $b(x,z)$ is called regular if it is a continuous, differentiable, strictly increasing and concave function of z, $b(x,0)=0$ and $b(x,\infty)=\beta \leq 1$. An important case is the exponential detection function, $b(x,z) = 1 - \exp(-\alpha(x)z)$. Sometimes this is called the detection function of random search.

If the detection information has the possibility of false contact, the search problem is called a noisy problem. In this case, the search operation usually consists of two stages, a broad search and an investigating search. The broad search is a search of preliminary nature, where contacts indicating possible target positions are coming in sequentially. The contacts gained in the broad search must be investigated in the second stage of the search (the investigating search) to ascertain whether the contact is true or not.

In evaluating the effectiveness of the search plan, several viewpoints are possible. The following measures of effectiveness are the most common among the objectives of optimization employed in search theory:

1. To maximize the detection probability of the target during the search.
2. To minimize the expected time until the detection of the target.
3. To minimize the expected cost until the detection of the target.
4. To maximize the whereabouts probability; this is defined as the probability of correctly stating the target position by either detecting the target or guessing its position at the end of an unsuccessful search.
5. To maximize the expected entropy gain.
6. To minimize the expected risk; this is defined as the expected searching cost until detection or stopping, whichever comes first, subtracted by the expected reward gained when the target is detected. Sometimes, a negative expected risk is named the expected return and is adopted as the measure to be maximized.

The block diagram shown in Fig. 1 presents the various attributes of search models in modeling the search operation. A set of assumptions selected by following down the block diagram from top to bottom specifies a search model.

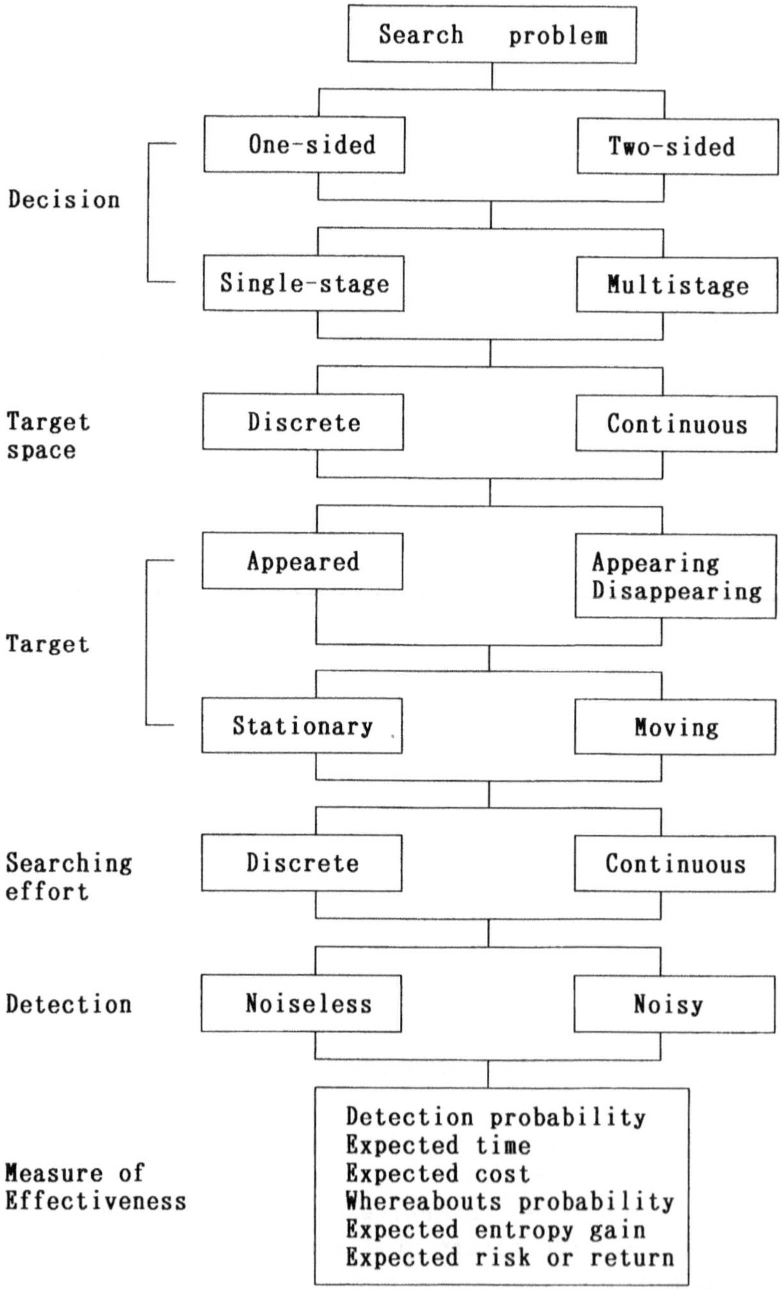

Figure 1. Diagram of search problem attributes and assumptions.

Outline of Chapters

The rest of this monograph will adhere to the following outline. In Chapter 1, an optimal search plan for a stationary target, which minimizes the expected risk of the search, is investigated. The search problem dealt with in this chapter is characterized by the set of assumptions shown in Fig. 1: one-sided multistage search, discrete target space, appeared stationary target, continuous searching effort, noiseless, and expected risk. First, a brief survey of the previous studies on this problem is given in Section 1.1. Then we formulate the problem as a variational problem and derive the necessary and sufficient conditions for the optimal search plan in Section 1.2. The theorems that specify how to search and when to stop the search are presented and the properties of the optimal search plan are elucidated. In Section 1.3, the optimal search plan for the two-box case is examined in detail and the complete solution of this problem is given.

In Chapter 2, the optimal search plans for a moving target are investigated. The problems are specified by the following set of assumptions: one-sided multistage search, discrete or continuous target space, appeared and moving target, continuous searching effort, noiseless, and detection probability or expected risk. First, a brief survey of this problem is given in Section 2.1. Then we investigate the optimal search plan for a moving target that maximizes the detection probability in Section 2.2. The problem is formulated as a nonlinear programming problem and a necessary and sufficient condition for the optimal distribution of searching effort is derived. In Section 2.3, we study the optimal search plan for a moving target that minimizes the expected risk. The problem is formulated as a variational problem. The necessary conditions for the optimal allocation of searching effort and the optimal stopping time of the search are obtained. Then we deal with a moving target with a rather simple type of motion, the factorable conditionally deterministic motion (F.C.D.M.) in Section 2.4. The relation between the optimal search plan maximizing the detection probability uniformly and the plan minimizing the expected risk is established. Using this relation, we derive the necessary and sufficient conditions for the optimal search plan for the F.C.D.M. target moving diffusively from a fixed point. We also derive the optimal search plans explicitly for four cases of the target speed distribution (the uniform, triangular, Rayleigh, and gamma distributions).

In Chapter 3, we derive an optimal whereabouts search plan minimizing the expected risk. The problems are characterized by the following set of assumptions: one-sided multistage search, discrete target space, appeared or disappearing stationary target, continuous searching effort, noiseless, and expected risk. First a brief survey of previous studies on the whereabouts search model is given in Section 3.1. Then, in Section 3.2, we investigate the optimal whereabouts search plan for an appeared stationary target; this specifies how to search, when to stop the search, and where a guess is to be made. We obtain a necessary and sufficient condition for the optimal distribution of searching effort and necessary conditions for the optimal stopping time of the search and for the optimal box to be guessed. The physical meanings of the theorems are also elucidated. In Section 3.3, the model described in the preceding section is generalized and a whereabouts search model for a disappearing (mortal) target is considered. The conditions for the optimal search and guess are derived and the effect of the lifetime of the target on the optimal search plan is discussed. Then, in Section 3.4, we examine several examples of the whereabouts search to see the influence of the lifetime of the target on the optimal plan.

In Chapter 4, we study an optimal investigating search plan for contacts in a two-stage search. We consider a problem minimizing the expected time of detecting the true target when noise-type false contacts are possible. The assumptions are: one-sided two-stage search, continuous target space, appeared moving target, continuous searching effort, noisy, and expected time. First, we briefly survey previous studies on the two-stage search problem in Section 4.1. Then, in Section 4.2, the problem is formulated as a fractional programming problem; the necessary and sufficient conditions for both the optimal selection of contacts to be investigated and the optimal stopping time of the investigating search for the selected contacts are derived. The properties of the optimal plan are also elucidated. Finally, several numerical examples are presented and distinctive features of the optimal investigating search plan are discussed in Section 4.3.

In Chapter 5, we analyze a forestalling detection problem in a two-sided search. The problems are specified by the following assumptions: two-sided single-stage search, continuous target space, appearing and disappearing stationary or moving target, continuous searching effort, noiseless, and detection probability and other characteristics of the search process. First a brief introduction to the problems dealt with in this chapter is given. Then we investigate a broad search process with detections forestalled by the target in Section 5.2. The problem is formulated as a continuous-time Markov chain model with five states, including two absorbing states (the detected target state and the disappeared target state). Formulas of the characteristics of the search process, such as the renewal density function, the expected number of transitions, the state distribution, and the expected time of absorption are derived explicitly. Using these formulas, a search problem in which the search time is constrained is formulated and a necessary condition for the optimal starting time of the search that maximizes the detection probability of the target is derived. Finally, we deal with a barrier search problem with forestalling by the target in Section 5.3. A relation between the detection probability of the target in a one-sided search and the forestalling detection probability in a two-sided search is obtained. Applying the relation, we derive the forestalling detection probability for such patterns of the barrier search as the fixed-point barrier search, the crossover barrier patrol, the back-and-forth barrier patrol, and the random area search including the barrier line.

CHAPTER 1

Optimal Search Plan for a Stationary Target Minimizing the Expected Risk

1.1. Introduction

In this chapter, we investigate an optimal search plan for a stationary target that minimizes the expected risk with a definite amount of continuous searching effort per unit time.

The study of the optimal distribution of searching effort for a stationary target was begun by Koopman (1957). Assuming the exponential detection function, he considered a problem maximizing the probability of detecting the target and showed how to allocate a given amount of continuous searching effort into the continuous target space. Charnes and Cooper (1958) established an algorithm for numerical calculation of the optimal distribution of searching effort given by Koopman, and de Guenin (1961) dealt with a generalized version of Koopman's model with the regular detection function. Staroverov (1963), Matula (1964), and Kadane (1969) investigated discrete versions of this model. The optimal distribution of searching effort minimizing the expected time until detection was first studied by Gilbert (1959). He considered a two-box search model with loss time imposed whenever the search region is switched. This model was generalized by Kisi (1966). Gluss (1961) dealt with a special n-box problem of this type in which the boxes are arranged in a row and Onaga (1971) investigated an n-box model with constant loss time. Lössner and Wegener (1982) considered a discrete search model with switching loss, and obtained conditions for the optimal search minimizing the expected searching cost until detection. As for the optimal search problem in which switching loss is not imposed, Dobbie (1963) and Black (1965) established the relation between the optimal distribution of searching effort maximizing the detection probability and the optimal effort distribution minimizing the expected time until detection.

In the search models mentioned here, the search is always started and is stopped when the target is detected or the available searching effort is exhausted. However, in some classes of search, whether the search should be started and when to stop if it is started are also important problems to be investigated. In order to analyze these problems, the expected risk is employed for the measure of effectiveness of a search plan. The optimal search plan minimizing the expected risk was first studied by Chew (1967, 1973) and later by Ross (1969) in a more general form. Ross dealt with the problem in a discrete effort model and formulated it by use of dynamic programming. He obtained several conditions for the optimal search plan that minimize the expected risk and discussed some approximations to the optimal search plan. Kan (1974) dealt with a special two-box case of Ross' problem. A continuous-effort version of Ross' model was investigated by Iida and theorems of the optimal search plan in a two-box problem (Iida 1982) and an n-box problem (Iida 1984) were obtained. Richardson and Belkin (1972) studied the stopping rule for continuous search, assuming a continuous target space. They assumed the homogeneous target space with respect to both searching cost and reward. Kisi and Tatsuno (1981) and Tatsuno and Kisi (1982) investigated a search and stop problem in which the capability of the detection device and the existence of the target in the target space are not known with certainty. They evaluated the effectiveness of using "dummies" to obtain extra information that gives a better estimate of the posterior distribution of the detection capability. In these studies, the detection capability is not varied in time, whereas it is unknown to the searcher. Nakai (1982) investigated an optimal search for a target with a random lifetime. Assuming the homogeneity of searching cost in both the target space and the time space, he obtained a necessary and sufficient condition for the optimal distribution of searching effort and a necessary condition for the optimal stopping time. Nakai (1987) also considered a model in which the searcher can improve his detection rate. In his model, it is assumed that the searcher allocates his time not only to search the target but also to improve the detection rate of his detection device. A necessary condition for the optimal allocation of time maximizing the detection probability of the target is obtained.

In this chapter, we deal with a problem in which the searching cost and reward vary over the target space but do not depend on time, the capability of the detection device does not vary in time and is known to the searcher, and the target is assumed to be immortal and stationary. In the next section, the optimal distribution of searching effort and the optimal stopping time of the search are investigated and properties of the optimal search plan are elucidated. (This section is based on Iida 1984.) Then, in Section 1.3, we examine the two-box case in detail and show the complete structure of the optimal search plan. (This section is taken from Iida 1982.)

1.2. Optimal Search Plan

1.2.1. Optimal Distribution of Searching Effort

In this section, we assume that the target space consists of n boxes and an appeared stationary target is known to be in one of the boxes with a probability vector $\mathbf{p} = \{p_i\}$, where p_i is the initial probability of the target being in box i, $i = 1, 2, \cdots, n$, $\Sigma_i p_i = 1$. Suppose that the target is not detected until t. Let $\mathbf{p}(t)$ be the posterior probability vector of the target distribution. We say that the state of the search process is in $\mathbf{p}(t)$ at t. We also assume that the search is carried out continuously in time subject to the restriction of a definite amount of total searching cost C per unit time. The total searching cost rate C is assumed to be divisible arbitrarily in allocating it to boxes, and search in box i is assumed to cost c_i (>0) per unit density of searching effort. The searcher is given a reward R_i (>0) when he successfully detects the target in box i. The conditional probability of detecting the target in box i with unit density of searching effort is assumed to be α_i (>0), irrespective of the history of the search. This assumption implies that the conditional detection function follows the random search formula; $1 - \exp(-\alpha_i \psi(i, t))$, where $\psi(i, t)$ is the cumulative density of searching effort applied to box i until t. Let T be a stopping time of the search and $\phi_T(i, t)$ be the density of searching effort per unit time allocated to box i at t. Then obviously $\phi_T(i, t) = \partial \psi_T(i, t)/\partial t$. A set $\boldsymbol{\phi}_T = \{\phi_T(i, t), i=1, 2, \cdots, n, 0 \leq t \leq T\}$ is called the search plan, which specifies both the distribution of searching effort and the stopping time of the search. The starting time and stopping time of the search in box i are denoted by T_i^s and T_i^e, respectively. The search plan that minimizes the expected risk is called the optimal search plan and is denoted by $\boldsymbol{\phi}_{T*}^* = \{\phi_{T*}^*(i, t), i=1, 2, \cdots, n, 0 \leq t \leq T^*\}$. Parameters with the superscript *, for example, T^*, also denote the optimal value.

If the reward R_i is sufficiently large, we should continue searching until detection of the target. But in some classes of search, the searching cost is relatively expensive compared with the reward. In this case, attention must be paid to the problem of stopping the search before detection. In order to analyze this problem, the expected risk of the search is adopted as the measure of effectiveness of a search plan. The definition of expected risk is the expected searching cost until detection or stopping, whichever comes first, minus the expected reward gained by detection of the target. Considering the situation where the state is in $\mathbf{p}$ at $t = 0$ and a search plan $\boldsymbol{\phi}_T$ is employed until T, the expected risk $f(\boldsymbol{\phi}_T | \mathbf{p})$ is written as follows:

$$f(\boldsymbol{\phi}_T|\,\mathbf{p}) = \Sigma_i p_i[\int_0^T \alpha_i \phi_T(i,t)\{\Sigma_j c_j \psi_T(j,t)-R_i\} \times \exp(-\alpha_i \psi_T(i,t))dt + \{\Sigma_j c_j \psi_T(j,T)\}\exp(-\alpha_i \psi_T(i,T))].$$

Since $\phi_T(i,t)$ is the density of searching effort per unit time allocated to box i at t, the constraint,

$$\Sigma_i c_i \phi_T(i,t) \leqq C \quad \text{for all } t \in [0,T],$$

is imposed. However, in this problem, there is an optimal search plan in which the searcher uses all the searching cost rate C exhaustively at each time, since the target is assumed to be stationary and both the reward and the searching cost do not vary in time. Therefore, the inequality in the constraint can be replaced by an equality and the constraint $\Sigma_i c_i \phi_T(i,t) = C$ is imposed for all $t \in [0,T]$. Substituting this relation into $f(\boldsymbol{\phi}_T|\,\mathbf{p})$, we obtain the following simplified expression:

(1.1) $$f(\boldsymbol{\phi}_T|\,\mathbf{p}) = \Sigma_i p_i[C\int_0^T \exp(-\alpha_i \psi_T(i,t))dt - R_i\{1-\exp(-\alpha_i \psi_T(i,T))\}] = \Sigma_i p_i[\int_0^T\{C - \alpha_i \phi_T(i,t)R_i\}\exp(-\alpha_i \psi_T(i,t))dt].$$

Hence, the problem is formulated as a variational problem to find the optimal functions $\{\phi_{T^*}{}^*(i,t)\}$ and the optimal stopping time T^* that minimize the functional $f(\boldsymbol{\phi}_T|\,\mathbf{p})$ subject to the following constraints:

(1.2a) $$\Sigma_i c_i \phi_T(i,t) = C \quad \text{for all } t \in [0,T],$$
(1.2b) $$\phi_T(i,t) \geqq 0 \quad \text{for all } i \text{ and } t \in [0,T].$$

The search plan that satisfies these constraints for all ts is called a feasible search plan and the set of the feasible search plan is denoted by $\boldsymbol{\Psi}_0$.

If the stopping time T is given in advance of the search, the search plan that minimizes the expected risk given by Eq.(1.1) is called the conditionally optimal search plan and is denoted by $\boldsymbol{\phi}_T^* = \{\phi_T^*(i,t)\}$ (and similarly $\psi_T^*(i,t)$). The following theorem gives the conditionallyoptimal search plan. (From now on, if there is no confusion, the expected risk $f(\boldsymbol{\phi}_T|\,\mathbf{p})$ is abbreviated as $f(\boldsymbol{\phi}_T)$ and the subscript T in ϕ_T and ψ_T is omitted.)

Theorem 1.1. *A necessary and sufficient condition for* $\{\phi^*(i,t)\}$ *to be conditionally optimal is*:
if $\phi^*(i,t) > 0$,

(1.3a) $$\frac{\alpha_i p_i}{c_i}[R_i\exp(-\alpha_i \psi^*(i,T)) + C\int_t^T \exp(-\alpha_i \psi^*(i,z))dz] = \lambda(t),$$

if $\phi^*(i,t) = 0$,

(1.3b) $$\frac{\alpha_i p_i}{c_i}[R_i\exp(-\alpha_i \psi^*(i,T)) + C\int_t^T \exp(-\alpha_i \psi^*(i,z))dz] \leqq \lambda(t),$$

for all i and a.e. t in $[0, T]$. *Here,* $\lambda(t)$ *is a nonnegative function and is determined by the constraint* $\Sigma_i c_i \phi(i, t) = C$.

<u>Proof.</u> The proof of the necessity is given by considering the variation of $\phi^*(i, t)$ in both time and boxes in Ψ_0. If $\phi^*(i, t)$ is dense at t, the conditional detection probability in the $\varepsilon/2$ neighborhood of t is zero from the definition of the detection function. Therefore, the conditionally optimal search plan $\boldsymbol{\phi}^*$ is not always dense in any time interval and we can find box i such as $\phi^*(i, t) > 0$ in $[t_1 - \varepsilon/2,\ t_1 + \varepsilon/2]$ for any t_1 ($<T$). We consider the search plan $\boldsymbol{\phi}$ which differs from $\boldsymbol{\phi}^*$ only in the $\varepsilon/2$ neighborhood of t_1,

if $|t - t_1| \leqq \varepsilon/2$,

$$\begin{aligned} \phi(i, t) &= \phi^*(i, t) - e/c_i, \\ \phi(j, t) &= \phi^*(j, t) + e/c_j, \\ \phi(k, t) &= \phi^*(k, t) \quad \text{for all } k \neq i, j, \end{aligned} \tag{1.4a}$$

if $|t - t_1| > \varepsilon/2$,

$$\phi(i, t) = \phi^*(i, t) \quad \text{for all } i. \tag{1.4b}$$

The allocation $\boldsymbol{\phi}$ is obtained from $\boldsymbol{\phi}^*$ by transferring some small searching budget e (>0) from box i to some box j in the time interval $[t_1 - \varepsilon/2,\ t_1 + \varepsilon/2]$. Since $\phi^*(i, t) > 0$ in this interval, this transfer is always possible if e is sufficiently small and $\boldsymbol{\phi}$ satisfies the constraints in Eqs. (1.2a) and (1.2b). From Eqs. (1.1), (1.4a) and (1.4b), we obtain

$$\begin{aligned} f(\boldsymbol{\phi}) - f(\boldsymbol{\phi}^*) &= p_i\{\exp(\alpha_i e\varepsilon/c_i) - 1\}[R_i \exp(-\alpha_i \phi^*(i, T)) \\ &\qquad + C\int_{t_1}^{T} \exp(-\alpha_i \phi^*(i, z))\,dz] \\ &\quad + p_j\{\exp(-\alpha_j e\varepsilon/c_j) - 1\}[R_j \exp(-\alpha_j \phi^*(j, T)) \\ &\qquad + C\int_{t_1}^{T} \exp(-\alpha_j \phi^*(j, z))\,dz] \geqq 0. \end{aligned}$$

If εe is sufficiently small, this relation can be written as

$$\begin{aligned} &\frac{\alpha_i p_i}{c_i}[R_i \exp(-\alpha_i \phi^*(i, T)) + C\int_{t_1}^{T} \exp(-\alpha_i \phi^*(i, z))\,dz] \\ &\qquad \geqq \frac{\alpha_j p_j}{c_j}[R_j \exp(-\alpha_j \phi^*(j, T)) + C\int_{t_1}^{T} \exp(-\alpha_j \phi^*(j, z))\,dz]. \end{aligned}$$

Here, we assume $\phi^*(j, t) > 0$ in this interval and consider a transfer of small searching cost e from box j to i. Then we obtain a relation with the opposite inequality sign. Therefore,

$$\frac{\alpha_i p_i}{c_i}[R_i \exp(-\alpha_i \phi^*(i, T)) + C\int_{t}^{T} \exp(-\alpha_i \phi^*(i, z))\,dz] = \lambda(t)$$

for all i such that $\phi^*(i, t) > 0$. And if $\phi^*(i, t) = 0$ and $\phi^*(j, t) > 0$, then

$$\frac{\alpha_i p_i}{c_i}[R_i \exp(-\alpha_i \phi^*(i, T)) + C\int_{t}^{T} \exp(-\alpha_i \phi^*(i, z))\,dz] \leqq \lambda(t).$$

To prove the sufficiency, the convexity of $f(\boldsymbol{\phi})$ with respect to $\boldsymbol{\phi}$ is proved easily by the convexity of the exponential function:

$$f(\boldsymbol{\phi}) \leqq (1-\theta)f(\boldsymbol{\phi}^1) + \theta f(\boldsymbol{\phi}^2),$$

where $\boldsymbol{\phi}^1$ and $\boldsymbol{\phi}^2$ are arbitrary search plans in $\boldsymbol{\Psi}_0$ and

$$\boldsymbol{\phi} = (1-\theta)\boldsymbol{\phi}^1 + \theta\boldsymbol{\phi}^2, \qquad 0 \leqq \theta \leqq 1.$$

Suppose that $\boldsymbol{\phi}^*$ is a search plan that satisfies Eqs.(1.3a) and (1.3b), and $\boldsymbol{\phi}^0$ is an arbitrary search plan in $\boldsymbol{\Psi}_0$. Considering a sufficiently small ε for θ and setting $\boldsymbol{\phi}^1 = \boldsymbol{\phi}^*$ and $\boldsymbol{\phi}^2 = \boldsymbol{\phi}^0$ in the preceding relation, we obtain

$$\begin{aligned}&\varepsilon\{f(\boldsymbol{\phi}^*)-f(\boldsymbol{\phi}^0)\}\\ &\qquad \leqq \varepsilon\{\int_0^T \lambda(t)[\Sigma_i c_i \phi^0(i,t)-\Sigma_i c_i \phi^*(i,t)]dt\} = 0.\end{aligned}$$

Therefore, the search plan satisfying Eqs.(1.3a) and (1.3b) is proved optimal since the relation $f(\boldsymbol{\phi}^*) \leqq f(\boldsymbol{\phi}^0)$ holds for arbitrary search plan $\boldsymbol{\phi}^0$. □

<u>Remark 1.1.</u> Let us consider the physical meaning of the preceding theorem. Denoting the searching effort allocated to box i in a time interval $(t,s]$ as

$$\zeta^*(i,t,s) = \int_t^s \phi^*(i,z)dz,$$

we can rewrite the left-hand side of Eqs.(1.3a) and (1.3b) as follows.

$$(1.5) \qquad \frac{\alpha_i p_i \exp(-\alpha_i \phi^*(i,t))}{c_i} \times [R_i + \int_t^T \exp(-\alpha_i \zeta^*(i,t,s))\{C-\alpha_i \phi^*(i,s)R_i\}ds].$$

The second term in the bracket,

$$\int_t^T \exp(-\alpha_i \zeta^*(i,t,s))\{C-\alpha_i \phi^*(i,s)R_i\}ds,$$

is the expected risk of the optimal search from t to T, when the target is in box i and is not detected until t. Here we recall that the search is to be continued until T if the target is not detected. Therefore, if we detect the target in box i at t, we earn the reward R_i and save the risk of the search, which is to be incurred in $(t,T]$ if the target is not detected. The sum of R_i and the saved risk can be interpreted as the return that motivates the search in box i at t. Meanwhile, the numerator in the first term of Eq.(1.5), $\alpha_i p_i \exp(-\alpha_i \phi^*(i,t))$, is the detection probability of the target with the unit density of searching effort allocated in box i at t and the denominator, c_i, is the cost of the unit density of searching effort. Hence, the left-hand side of Eqs.(1.3a) and (1.3b) means the expected marginal return versus cost ratio with the unit density of searching effort allocated to box i at t. Therefore, the theorem means that "if the searching effort is to be allocated to box i at t, the density of the searching effort should be determined in such a way that the expected marginal return versus cost ratio is balanced to $\lambda(t)$ among the boxes being searched at t, and if the searching effort should not be allocated to box i, the value of the ratio in box i is smaller than or equal to $\lambda(t)$."

Remark 1.2. We obtain the theorem under the assumptions of the exponential detection function, the discrete target space, and the continuous time space. However, these assumptions are not essential to the problem. The model is generalized easily by replacing the exponential detection function with a regular detection function. The assumption of the discrete target space can be changed with the continuous target space without any difficulty and the similar theorem is easily obtained. The continuous time space is also changed easily with the discrete time point system. In this section, we have assumed that the total searching cost rate C can be divided arbitrarily in allocating it to boxes. This assumption is very important for our model. If the searching cost rate is quantized, the condition for the optimal search plan would be complicated drastically.

The following corollary is obtained directly by setting $t = T$ in the theorem.

Corollary 1.2. *If* $\phi^*(i,T) > 0$,

$$\frac{\alpha_i p_i}{c_i} R_i \exp(-\alpha_i \phi^*(i,T)) = \lambda(T),$$

if $\phi^*(i,T) = 0$,

$$\frac{\alpha_i p_i}{c_i} R_i \exp(-\alpha_i \phi^*(i,T)) \leqq \lambda(T).$$

The function $\lambda(t)$ in the theorem has the following properties.

Corollary 1.3. *The function $\lambda(t)$ is a continuous, strictly decreasing and strictly convex function of t.*

Proof. For arbitrary t and s (>0), $t+s \leqq T$, we can always find box i and j such that $\phi^*(i,t) > 0$ and $\phi^*(j,t+s) > 0$. Then we have this relation:

$$C\frac{\alpha_j p_j}{c_j}\int_t^{t+s} \exp(-\alpha_j \phi^*(j,z))\,dz \leqq \lambda(t) - \lambda(t+s)$$
$$\leqq C\frac{\alpha_i p_i}{c_i}\int_t^{t+s} \exp(-\alpha_i \phi^*(i,z))\,dz.$$

Therefore, $\lim_{s\to 0}\lambda(t+s) = \lambda(t)$ and $\lambda(t) > \lambda(t+s)$; i.e., $\lambda(t)$ is continuous and strictly decreasing. The convexity of $\lambda(t)$ is proved as follows. Consider time points t_1 and t_2, $t_1 < t_2 \leq T$, and $t = (1-\theta)t_1 + \theta t_2$, $0 \leq \theta \leq 1$. We can always find box i such that $\phi^*(i,t) > 0$, and from Theorem 1.1, we have

$$\lambda(t) = \frac{\alpha_i p_i}{c_i}[R_i \exp(-\alpha_i \phi^*(i,T)) + C\int_t^T \exp(-\alpha_i \phi^*(i,z))dz]$$
$$\leq (1-\theta)\lambda(t_1) + \theta\lambda(t_2) + \frac{\alpha_i p_i}{c_i} C[\theta \int_t^{t2} \exp(-\alpha_i \phi^*(i,z))dz$$
$$- (1-\theta)\int_{t1}^{t} \exp(-\alpha_i \phi^*(i,z))dz].$$

Since $\phi^*(i,z)$ is a nondecreasing function of z, the last term in the brackets of the above relation is nonpositive. Therefore, a relation $\lambda(t) \leq (1-\theta)\lambda(t_1) + \theta\lambda(t_2)$ is established. Since the equality holds only for $\theta=0$ or $\theta=1$, $\lambda(t)$ is a strictly convex function of t. □

Let $\mathbf{D}(t)$ and $\mathbf{S}(t)$ be the sets of boxes that have been searched until t and that are being searched at t, respectively,

(1.6a) $$\mathbf{D}(t) = \{i | \phi^*(i,t) > 0\},$$
(1.6b) $$\mathbf{S}(t) = \{i | \phi^*(i,t) > 0\}.$$

The following theorem elucidates the properties of the conditionally optimal search plan.

<u>Theorem 1.4.</u> *Suppose* $0 \leq t_1 < t_2 \leq T$ *and* $\phi^*(i,t)$ *is not dense at* t_1 *and* t_2.

1. *If* $\phi^*(i,t)$ *is positive in the interval* $[t_1, t_2)$ *and is zero at* t_2, $\phi^*(i,t)$ *is always zero throughout the interval* $[t_2, T]$.

2. *If* $\phi^*(i,t)$ *is positive at both* t_1 *and* t_2, *then* $\phi^*(i,t)$ *is always positive in the interval* $[t_1, t_2]$.

3. *If both* $\phi^*(i,t)$ *and* $\phi^*(j,t)$ *are positive in the interval* $[t_1, t_2]$, *then at any time* t *in* $[t_1, t_2]$,

(1.7) $$\frac{\alpha_i p_i}{c_i}\exp(-\alpha_i \phi^*(i,t)) = \frac{\alpha_j p_j}{c_j}\exp(-\alpha_j \phi^*(j,t)),$$

and

(1.8) $$\alpha_i \phi^*(i,t) = \alpha_j \phi^*(j,t) = \frac{C}{\Sigma_{S(t)}\{c_k/\alpha_k\}}.$$

<u>Proof.</u> 1. We assume $\phi^*(i,t) > 0$ in some $t \in (t_2, T]$. Let t_3 be the smallest time point in $t \in (t_2, T]$ at which $\phi^*(i,t_3) > 0$. Applying Theorem 1.1, we obtain the following relation at some t, $t_2 < t < t_3$,

$$\frac{\lambda(t_3) - \lambda(t_2)}{t_3 - t_2} \geq \frac{\lambda(t_3) - \lambda(t)}{t_3 - t}.$$

This relation contradicts the strict convexity of $\lambda(t)$ and therefore, we can conclude $\phi^*(i,t) = 0$ in $[t_2, T]$.

2. If we assume $\phi^*(i,t_3) = 0$ at some time point t_3, $t_1 < t_3 < t_2$, $\phi^*(i,t) = 0$ is concluded in the interval $[t_3, T]$ from the preceding

result. This contradicts the assumption, $\phi^*(i, t_2) > 0$.

3. From Theorem 1.1, we have

$$\lambda(t_3) - \lambda(t) = \frac{\alpha_i p_i}{c_i} C\int_{t_1}^{t} \exp(-\alpha_i \phi^*(i, z))\, dz$$
$$= \frac{\alpha_j p_j}{c_j} C\int_{t_1}^{t} \exp(-\alpha_j \phi^*(j, z))\, dz.$$

Since this relation holds at any t in $[t_1, t_2]$, we can conclude Eq. (1.7). Eq. (1.8) is derived easily from Eq. (1.7). □

Let T_i^s and T_i^e be the starting time and stopping time of search in box i, respectively. The following theorem is established.

Theorem 1.5. *Suppose both box i and j are the elements of* $\mathbf{D}(T)$.

1. *If and only if* $\alpha_i p_i / c_i \geq \alpha_j p_j / c_j$, *then* $T_i^{s*} \leq T_j^{s*}$.
2. *If and only if* $T_i^{e*} \geq T_j^{e*}$, *then*

$$\frac{\alpha_i p_i \exp(-\alpha_i \phi^*(i, T))}{c_i} \leq \frac{\alpha_j p_j \exp(-\alpha_j \phi^*(j, T))}{c_j}.$$

3. *If and only if* $R_i \geq R_j$, *then* $T_i^{e*} \geq T_j^{e*}$. *Therefore, if and only if box i is an element of* $\mathbf{S}(T)$, R_i *is the largest reward in* $\mathbf{D}(T)$.

Proof. 1. Applying Theorem 1.1 to box i and j at T_i^{s*} and T_j^{s*}, we obtain

$$(1.9) \qquad \frac{\alpha_i p_i}{c_i} \int_{T_j^{s*}}^{T_i^{s*}} \exp(-\alpha_i \phi^*(i, z))\, dz \leq \frac{\alpha_j p_j}{c_j} \int_{T_j^{s*}}^{T_i^{s*}} \exp(-\alpha_j \phi^*(j, z))\, dz.$$

Here we assume $T_i^{s*} > T_j^{s*}$. Noting $\phi^*(i, t) = 0$ and $\phi^*(j, t) > 0$ in any $t \in [T_j^{s*}, T_i^{s*})$ and using Eq. (1.9), we obtain $\alpha_i p_i / c_i < \alpha_j p_j / c_j$. Since this contradicts the assumption, we can conclude $T_i^{s*} \leq T_j^{s*}$.

To prove the sufficiency, if $T_i^{s*} \leq T_j^{s*}$, then $\phi^*(i, t) > 0$ and $\phi^*(j, t) = 0$ in $t \in [T_i^{s*}, T_j^{s*})$, and therefore we have $\alpha_i p_i / c_i \geq \alpha_j p_j / c_j$ from Eq. (1.9).

2. Considering that $\phi^*(i, t)$ is a nondecreasing function of t and both $\phi^*(i, T_i^{e*})$ and $\phi^*(j, T_j^{e*})$ are positive, we obtain from Theorem 1.1,

$$(1.10) \qquad \frac{\alpha_i p_i}{c_i} \int_{T_j^{e*}}^{T_i^{e*}} \exp(-\alpha_i \phi^*(i, z))\, dz \leq \frac{\alpha_j p_j}{c_j} \int_{T_j^{e*}}^{T_i^{e*}} \exp(-\alpha_j \phi^*(j, z))\, dz,$$

$$\frac{\alpha_i p_i}{c_i} \exp(-\alpha_i \phi^*(i, T)) \leq \frac{\alpha_j p_j}{c_j} \exp(-\alpha_j \phi^*(j, T)).$$

The sufficient condition is proved by use of the reductive absurdity. We assume $T_i^{e*} < T_j^{e*}$ under the condition

$$\frac{\alpha_i p_i}{c_i} \exp(-\alpha_i \phi^*(i,T)) \leqq \frac{\alpha_j p_j}{c_j} \exp(-\alpha_j \phi^*(j,T)),$$

then a relation contradicting Eq. (1.10) is derived.

3. To use the reductive absurdity, we assume $T_i^{e*} < T_j^{e*}$ under $R_i \geqq R_j$. From Theorem 1.5-2, we have

$$\frac{\alpha_i p_i}{c_i} \exp(-\alpha_i \phi^*(i,T)) > \frac{\alpha_j p_j}{c_j} \exp(-\alpha_j \phi^*(j,T)).$$

On the other hand, since $\phi^*(i, T_j^{e*}) = 0$ and $\phi^*(j, T_j^{e*}) > 0$ by the assumption, we obtain the following relation from Theorem 1.1,

$$\frac{\alpha_i p_i \exp(-\alpha_i \phi^*(i,T))}{c_i} [R_i + C(T - T_j^{e*})]$$
$$\leqq \frac{\alpha_j p_j \exp(-\alpha_j \phi^*(j,T))}{c_j} [R_j + C(T - T_j^{e*})].$$

From these inequalities, we have $R_i < R_j$, and it contradicts the condition.

The sufficiency is proved by showing $R_i \geqq R_j$ if $T_i^{e*} \geqq T_j^{e*}$. In this case, Eq. (1.10) holds, and from $\phi^*(i, T_i^{e*}) > 0$ and $\phi^*(j, T_i^{e*}) = 0$, we obtain

$$\frac{\alpha_i p_i \exp(-\alpha_i \phi^*(i,T_i^{e*}))}{c_i} [R_i + C(T - T_i^{e*})]$$
$$\geqq \frac{\alpha_j p_j \exp(-\alpha_j \phi^*(j,T_i^{e*}))}{c_j} [R_j + C(T - T_i^{e*})].$$

From these inequalities, we obtain $R_i \geqq R_j$. □

The widely known theorem (for example, Tada 1964,1965) for the optimal search plan in which the search should not be stopped until detection is derived easily from Theorem 1.1.

<u>Corollary 1.6.</u> *If the optimal search never stops searching until detection, then for all i and $t \in [T^0, \infty)$,*

$$\alpha_i \phi^*_\infty(i,t) = \frac{C}{\Sigma_k (c_k/\alpha_k)}. \tag{1.11}$$

The searching effort allocation given by Eq. (1.11) is called the balanced search. T^0 is given by

$$T^0 = \frac{1}{C} [\Sigma_i \frac{c_i}{\alpha_i} \{\log \frac{\alpha_i p_i / c_i}{\alpha_j p_j / c_j}\}], \tag{1.12}$$

where j is the box having the minimum value of $\alpha_i p_i / c_i$. The posterior probability distribution of the target is

$$\mathbf{p}^0 = \{p_i(t)\} = \{\frac{c_i/\alpha_i}{\Sigma_k (c_k/\alpha_k)}\}.$$

<u>Remark 1.3.</u> It is worthwhile to point out that the myopic search

plan is not optimal in this problem. This is confirmed by noting the fact that Eqs.(1.3a) and (1.3b) which determine the optimal distribution of searching effort at t, contain all the search plan $\boldsymbol{\phi}^*$ in $[0, T]$. This property strongly depends on the inhomogeneity of the reward in the target space. To see it, suppose the reward R_i are identical for all boxes, $R_i = R$. In this case, $T_i^{e*} = T$ for all boxes $i \in \mathbf{D}(T)$ from Theorem 1.5-3. Therefore, $\mathbf{D}(T) = \mathbf{S}(T)$ and

$$\frac{\alpha_i p_i \exp(-\alpha_i \phi^*(i, t))}{c_i} = \text{constant}$$

for all $i \in \mathbf{S}(t)$ from Theorem 1.4-3. Consequently, the conditions in Theorem 1.1 take the simple expression: if $\phi^*(i, t) > (=) 0$,

$$\frac{\alpha_i p_i R}{c_i} \exp(-\alpha_i \phi^*(i, t)) = (\leq) \lambda(t).$$

It should be noted that these relations do not depend on the search plan in $(t, T]$, and therefore, the optimal distribution of the searching effort is determined by the myopic search plan if $R_i = R$. This optimal search plan also maximizes the detection probability of the target at all time uniformly in $[0, T]$.

1.2.2. Optimal Stopping Time of the Search

In this section, we will investigate the optimal stopping time. For this purpose, we deal with the stopping time T as a variable. Let $\mathbf{p}(T)$ be the posterior probability vector of the target distribution given the conditionally optimal search plan until T fails to detect the target:

$$\mathbf{p}(T) = \{p_i(T)\} = \{ \frac{p_i \exp(-\alpha_i \phi_T^*(i, T))}{Q(T)} \},$$

where $Q(T)$ is the nondetection probability of the target with $\boldsymbol{\phi}^*$ and is given by $Q(T) = \Sigma_i p_i \exp(-\alpha_i \phi_T^*(i, T))$. Then, we have the following theorem.

<u>Theorem 1.7.</u> *A necessary condition for the optimal stopping time T^* is*

$$\begin{aligned} \lambda(T^*-\varepsilon) &\geq Q(T^*-\varepsilon), \\ \lambda(T^*) &= Q(T^*), \qquad (1.13) \\ \lambda(T^*+\varepsilon) &\leq Q(T^*+\varepsilon), \end{aligned}$$

where ε is a positive and sufficiently small time interval and $\lambda(T)$ is given by Corollary 1.2.

<u>Proof.</u> Let $\boldsymbol{\phi}_T^*$ be the conditionally optimal search plan with a stopping time T and $f_Z(\boldsymbol{\phi}_T^*)$ be the expected risk of the search in which the search plan $\boldsymbol{\phi}_T^*$ is stopped at Z $(<T)$. Suppose $T \leq T^*$. Since $f(\boldsymbol{\phi}_T^*)$ is a nonincreasing function of T in the neighborhood of

T^*, the following inequalities are obtained:

$$0 \geq f(\phi_T^*) - f(\phi_{T-\varepsilon}^*) \geq f(\phi_T^*) - f_{T-\varepsilon}(\phi_T^*)$$
$$= C\varepsilon[Q(T) - \lambda(T)] + o(\varepsilon).$$

Therefore, $Q(T) \leq \lambda(T)$ in the neighborhood of T^* if $T \leq T^*$.

Suppose $T \geq T^*$. The following relation is valid for $\varepsilon > 0$:

$$f(\phi_{T+\varepsilon}^* \mid \mathbf{p}) \leq f(\phi_T^* \mid \mathbf{p}) + Q(T) f(\phi_\varepsilon^* \mid \mathbf{p}(T)),$$

where ϕ_ε^* is the conditionally optimal search plan given that the initial state is $\mathbf{p}(T)$ and the duration of the search is ε. Then we have

$$Q(T) f(\phi_\varepsilon^* \mid \mathbf{p}(T)) = C\varepsilon[Q(T) - \lambda(T)] + o(\varepsilon).$$

Therefore, we have

$$0 \leq f(\phi_{T+\varepsilon}^* \mid \mathbf{p}) - f(\phi_T^* \mid \mathbf{p})$$
$$\leq C\varepsilon[Q(T) - \lambda(T)] + o(\varepsilon),$$

hence $Q(T) \geq \lambda(T)$ in $T \geq T^*$. From these relations, the theorem is established. □

It is obvious that the (unconditionally) optimal distribution of searching effort is determined by setting $T = T^*$ in Theorem 1.1 and $\phi_{T^*}^*$ is the optimal search plan.

By substituting $\lambda(T^*)$ given by Corollary 1.2 into Eq.(1.13), we can describe Theorem 1.7 in another way.

Corollary 1.8. *A necessary condition for the optimal stopping time T^* is*

$$\frac{\alpha_i p_i(T) R_i}{c_i} \geq 1 \quad \text{if } T < T^*,$$
$$\frac{\alpha_i p_i(T) R_i}{c_i} = 1 \quad \text{if } T = T^*,$$
$$\frac{\alpha_i p_i(T) R_i}{c_i} \leq 1 \quad \text{if } T > T^*,$$

for any T in the neighborhood of T^* and any i for which $\phi_T^*(i, T) > 0$.

The following corollary is derived directly from Corollary 1.8.

Corollary 1.9. 1. *A necessary condition for $T^* = 0$ is*

$$\max_i \frac{\alpha_i p_i R_i}{c_i} \leq 1.$$

2. *If $\max_i \alpha_i p_i(T) R_i / c_i > 1$, the search should not be stopped at T.*
3. *If $R_i \leq c_i / \alpha_i$ for all i, the search should not be started.*

The following theorem states the conditions for the case in which the search should not be stopped until detection.

Theorem 1.10. 1. *A necessary condition for* $T^* = \infty$ *is*

(1.14) $$\{\Sigma_i \frac{c_i}{\alpha_i}\}^2 \leqq \Sigma_i\{\frac{c_i R_i}{\alpha_i}\}.$$

2. *A sufficient condition for* $T^* = \infty$ *is*

$$\Sigma_i\{\frac{c_i}{\alpha_i R_i}\} \leqq 1.$$

Proof. 1. Suppose that the optimal search never stops until the target is detected. Then, as shown by Corollary 1.6, the optimal search $\phi_\infty^*(i,t)$ after T^0 is given by Eq.(1.11) and we have

$$p_i\exp(-\alpha_i\phi_\infty^*(i,t)) = \frac{c_i\lambda(T^0)\exp\{-C(t-T^0)/\Sigma_j(c_j/\alpha_j)\}}{\alpha_i\Sigma_j(c_j/\alpha_j)}$$

for all i and $t \geqq T^0$. Then we obtain

$$f(\phi_\infty^*) - f_{T0}(\phi_\infty^*) = \frac{\lambda(T^0)[\{\Sigma_i(c_i/\alpha_i)\}^2 - \Sigma_i(c_iR_i/\alpha_i)]}{\Sigma_i(c_i/\alpha_i)} \leqq 0.$$

Therefore, $\{\Sigma_i(c_i/\alpha_i)\}^2 \leqq \Sigma_i(c_iR_i/\alpha_i\}$ is established.

2. The condition is easily derived from Corollary 1.9-2. □

The following theorem states the properties of the boxes to be searched.

Theorem 1.11. *Let* $\mathbf{D}(T)$ *and* $\mathbf{S}(T)$ *be the sets defined by Eqs.* (1.6a) *and* (1.6b), *and suppose* $0 < T^* < \infty$.

1. *If* $\alpha_i p_i R_i/c_i > 1$, *box* i *is an element of* $\mathbf{D}(T^*)$.

2. *If* $i \in \mathbf{S}(T^*)$ *and* $\alpha_i p_i R_i/c_i < \alpha_j p_j R_j/c_j$, *then box* j *is an element of* $\mathbf{D}(T^*)$.

3. *If* $i \in \mathbf{D}(T^*)$ *and* $\alpha_i p_i R_i/c_i < \alpha_j p_j R_j/c_j$ *and* $\alpha_i p_i/c_i < \alpha_j p_j/c_j$, *then box* j *is also an element of* $\mathbf{D}(T^*)$.

4. *If* $i \in \mathbf{S}(T^*)$, R_i *is larger than* $\Sigma_{S(T*)} c_i/\alpha_i$.

Proof. 1. Since $\alpha_i p_i R_i/c_i > 1$, to stop immediately is not optimal by Corollary 1.9-2. Here we assume box i is not an element of $\mathbf{D}(T^*)$ and let box k be an element of $\mathbf{S}(T^*)$. From Corollary 1.2 and Theorem 1.7, we have

$$\frac{\alpha_i p_i R_i}{c_i} \leqq \frac{\alpha_k p_k R_k}{c_k}\exp(-\alpha_k\phi_{T*}^*(k,T^*))$$
$$= \lambda(T^*) = \Sigma_j p_j\exp(-\alpha_j\phi_{T*}^*(j,T^*)).$$

This implies $\Sigma_j p_j\exp(-\alpha_j\phi_{T*}^*(j,T^*)) > 1$ and is a contradiction. Therefore, $i \in \mathbf{D}(T^*)$ is concluded.

2. We assume box j is not an element of $\mathbf{D}(T^*)$. From Corollary 1.2, we obtain

$$\frac{\alpha_j p_j R_j}{c_j} \leqq \frac{\alpha_i p_i R_i}{c_i} \exp(-\alpha_i \phi_{T*}{}^*(i, T^*)) = \lambda(T^*).$$

Since $\alpha_j p_j R_j / c_j > \alpha_i p_i R_i / c_i$ by the assumption of the theorem, this inequality leads to a contradiction $\exp(-\alpha_i \phi_{T*}{}^*(i, T^*)) > 1$ for $\phi_{T*}{}^*(i, T^*) > 0$.

3. We assume that box i is an element of $\mathbf{S}(t)$ and box j is not an element of $\mathbf{D}(T^*)$. Applying Theorem 1.1, we have

$$\frac{\alpha_j p_j R_j}{c_j} - \frac{\alpha_i p_i R_i}{c_i} \exp(-\alpha_i \phi_{T*}{}^*(i, T^*))$$
$$\leqq C\int_t^{T*} \{ \frac{\alpha_i p_i}{c_i} \exp(-\alpha_i \phi_{T*}{}^*(i, z)) - \frac{\alpha_j p_j}{c_j} \} dz.$$

Since the left-hand side of this inequality is positive and the right-hand side is negative by the conditions of the theorem, this is a contradiction.

4. If $i \in \mathbf{S}(T^*)$, $\alpha_i p_i(T^*) R_i / c_i = 1$ by Corollary 1.8 and R_i is the same for all $i \in \mathbf{S}(T^*)$ by Corollary 1.5-3. Then we have

$$\Sigma_{S(T*)} \frac{c_j}{\alpha_j R_j} = \frac{1}{R_i} \Sigma_{S(T*)} \frac{c_j}{\alpha_j} = \Sigma_{S(T*)}\, p_j(T^*) < 1.$$

Therefore, if $i \in \mathbf{S}(T^*)$, then $R_i > \Sigma_{S(T*)} (c_j / \alpha_j)$. □

Remark 1.4. Theorem 1.7 and Corollary 1.8 state the behavior of $\lambda(T)$ at the neighborhood of the optimal stopping time T^*. From Corollary 1.3, $\lambda(t)$ is a strictly decreasing function of t and it becomes the minimum value at the stopping time T. If $\lambda(T) > Q(T)$, the optimal stopping time T^* is larger than T from Theorem 1.7, and $\lambda(T^*) = Q(T^*)$ at the optimal stopping time. The physical meaning of the condition for the optimal stopping time is elucidated from Corollary 1.8 as follows: The largest value of the expected marginal return versus cost ratio, $\alpha_i p_i(T) R_i / c_i$, decreases across unity as the stopping time T increases across T^*. Therefore, the search should be stopped when the expected marginal return becomes smaller than the cost.

Remark 1.5. Here we consider the meaning of Eq. (1.14) in Theorem 1.10. From Corollary 1.6, if $T^* = \infty$, the optimal search after T^0 is to search all boxes in parallel with the balanced effort distribution given by Eq. (1.11) and the posterior distribution of the target remains unchanged at

$$\mathbf{p}^0 = \{ \frac{c_i / \alpha_i}{\Sigma_j (c_j / \alpha_j)} \}.$$

Let $E(C|\,\mathbf{p}^0)$ be the expected searching cost until detecting the target given that the initial state is $\mathbf{p}^0$ and the optimal search given by Eq. (1.11) is employed. Let $E(R|\,\mathbf{p}^0)$ be the expected reward. Then from a simple recurrence relation, we obtain $E(C|\,\mathbf{p}^0) = \Sigma_i (c_i / \alpha_i)$ and $E(R|\,\mathbf{p}^0) = \Sigma_i p_i{}^0 R_i$. On the other hand, substituting $\mathbf{p}^0$ into Eq. (1.14), we have $\Sigma_i(c_i/\alpha_i) < \Sigma_i p_i^0 R_i$. Therefore, the condition in Eq.(1.14) means $E(C|\mathbf{p}^0) < E(R|\mathbf{p}^0)$, and in this case, it is reasonable that the search should not be stopped until detection.

1.3. Two-Box Problem

In this section, we examine the optimal search plan for a target in one of two boxes and derive the optimal search plan and the expected risk explicitly.

Applying Theorem 1.4 and Corollary 1.6 to the two-box problem, we can easily find that if the optimal stopping time is finite, the optimal search plan has a type

$$\boldsymbol{\phi}^* = \{\phi^*(i,t)=C/c_i,\ \ \phi^*(j,t)=0,\ \ 0\leq t\leq T<\infty\}.$$

That is, only one box is searched and the other box is left unsearched. From this, we can conclude that the optimal search plan is one of the following:

1. Search is stopped immediately (this plan is denoted by S_0).
2. Search is started in box i, $i=1$ or 2, and is continued until a finite stopping time T_i^e if the target is not found, and another box is never searched (this plan is denoted as S_i).
3. Search is started in box i and continued until T_i^0, two boxes are searched in parallel with the effort distribution given by Eq.(1.11) (denoted as S_i^0).

The expected risk for each search plan is obtained as follows:

Corollary 1.12. 1. *The expected risk for* S_0 *is obviously* $f(S_0)=0$.

2. *The expected risk for* S_i *is*

$$f(S_i) = \frac{c_i}{\alpha_i} - p_iR_i + (1-p_i)\frac{c_i}{\alpha_i}\log\left[\frac{\alpha_i p_i}{c_i(1-p_i)}\left(R_i-\frac{c_i}{\alpha_i}\right)\right],$$

where

$$R_i > \frac{c_i}{\alpha_i} \quad \text{and} \quad p_i > \frac{c_i}{\alpha_i R_i}. \tag{1.15}$$

The conditionally optimal stopping time T_i^e *in* S_i *is*

$$T_i^e = \frac{c_i}{\alpha_i C}\log\left[\frac{\alpha_i p_i}{c_i(1-p_i)}\left(R_i-\frac{c_i}{\alpha_i}\right)\right], \tag{1.16}$$

and the posterior probability distribution of the target at T_i^e *is*

$$\mathbf{p}(T_i^e) = (p_i(T_i^e),\ p_j(T_i^e)) = \left(\frac{c_i}{\alpha_i R_i},\ 1-\frac{c_i}{\alpha_i R_i}\right).$$

If Eq.(1.15) *is not satisfied, the optimal stopping time* T_i^o *is zero, and consequently,* $f(S_i)=0$.

3. *We define*

$$p_i(\infty) = \frac{c_i/\alpha_i}{c_1/\alpha_1 + c_2/\alpha_2}.$$

If S_i^0 is optimal, we have the expected risk: if $p_i > p_i(\infty)$,

$$(1.17) \qquad f(S_i^0) = -(p_1R_1+p_2R_2) + \frac{c_i}{\alpha_i} + (1-p_i)\left[\frac{c_1}{\alpha_1} + \frac{c_2}{\alpha_2} + \frac{c_i}{\alpha_i}\log\frac{\alpha_i p_i c_j}{\alpha_j(1-p_i)c_i}\right],$$

and if $p_i \leqq p_i(\infty)$,

$$(1.18) \qquad f(S_i^0) = \frac{c_1}{\alpha_1} + \frac{c_2}{\alpha_2} - (p_1R_1+p_2R_2).$$

The conditionally optimal switching time T_i^0 to the balanced search is

$$(1.19) \qquad \begin{aligned} T_i^0 &= \frac{c_i}{\alpha_i C}\log\frac{\alpha_i p_i c_j}{\alpha_j(1-p_i)c_i} && \text{if } p_i > p_i(\infty), \\ T_i^0 &= 0 && \text{if } p_i \leqq p_i(\infty), \end{aligned}$$

and the posterior probability distribution of the target is

$$(1.20) \qquad \begin{aligned} p_i(T_i^0) &= p_i(\infty) && \text{if } T_i^0 > 0, \\ p_i(T_i^0) &= p_i && \text{if } T_i^0 = 0. \end{aligned}$$

Proof. The conditionally optimal stopping time T_i^e given by Eq. (1.16), and the starting time of the balanced search T_i^0 given by Eq. (1.19), are calculated from Corollaries 1.8 and 1.6, respectively. The conditionally optimal expected risk $f(\cdot)$ of the search plan mentioned is readily obtained by substituting $\boldsymbol{\phi}^*$, T_i^e, or T_i^0 into Eq. (1.1). □

The following corollary is obtained directly from Corollary 1.12.

Corollary 1.13. 1. *The search plan S_i exists only for $R_i > c_i/\alpha_i$ and $p_i(T_i^e) < p_i \leqq 1$. Then, $f(S_i|\mathbf{p})$ is a strictly decreasing and concave function of p_i and*

$$\begin{aligned} f(S_i|\mathbf{p}(T_i^e)) &= 0, \\ f(S_i|(p_i=1, p_j=0)) &= \frac{c_i}{\alpha_i} - R_i, \\ f'(S_i|\mathbf{p}(T_i^e)) &= 0, \\ f'(S_i|(p_i=1, p_j=0)) &= -\infty, \end{aligned}$$

where f' is the derivative of f with respect to p_i.

2. *We define a function $g(\mathbf{p}) = \min_i f(S_i^0|\mathbf{p})$. Then we have*

$$\begin{aligned} g(\mathbf{p}) &= f(S_i^0|\mathbf{p}) && \text{if } p_i \geqq p_i(\infty), \\ &= f(S_j^0|\mathbf{p}) && \text{if } p_i < p_i(\infty),\ i \neq j. \end{aligned}$$

$g(\mathbf{p})$ is a continuous and concave function of p_i and

$$g(\mathbf{p}(\infty)) = \frac{c_1}{\alpha_1} + \frac{c_2}{\alpha_2} = \frac{c_1R_1/\alpha_1 + c_2R_2/\alpha_2}{c_1/\alpha_1 + c_2/\alpha_2},$$

$$g((p_i=1, p_j=0)) = f(S_i^0|(p_i=1, p_j=0)) = \frac{c_i}{\alpha_i} - R_i,$$

$$g'(\mathbf{p}(\infty)) = R_j - R_i, \quad g'((p_i=1, p_j=0)) = -\infty,$$

where g' is the derivative of g with respect to p_i.

It must be noted that the expected risk $f(S_i^0|\mathbf{p})$ in $p_i \leqq p_i(\infty)$ given by Eq.(1.18), with $i = 1,2$, are always larger than the expected risk given by Eq.(1.17), with $i = 2,1$, respectively. Further, the two curves given by Eq.(1.17), namely, $f(S_i^0|\mathbf{p})$ in $p_i \geqq p_i(\infty)$ and $f(S_j^0|\mathbf{p})$ in $p_i < p_i(\infty)$ are connected continuously and smoothly at $\mathbf{p}(\infty)$. Therefore, the function $g(\mathbf{p})$ does not contain the expected risk given by Eq.(1.18).

Since the conditionally optimal risk is given by Corollary 1.12, the (unconditionally) optimal risk at a state $\mathbf{p}$ is obtained by

(1.21) $$f(\boldsymbol{\phi}^*|\mathbf{p}) = \min\,[f(S_0|\mathbf{p}),\ f(S_i|\mathbf{p}),\ f(S_i^0|\mathbf{p})].$$

Until now, the state variable is defined by the probability vector of the target distribution. Since there are only two boxes, the state can be represented by a scalar $p_i(t)$, $i = 1$ or 2, instead of the vector $\mathbf{p}(t)$. Here, consider the state variable p_i is varied from 0 to 1 for a fixed set of system parameters. The optimal plan determined by Eq. (1.21) is unchanged for some interval of p_i, but in passing a point, the optimal plan makes a change, e.g., from S_2^0 to S_1^0. In some cases, more changes will occur if p_i is varied further. An optimal plan region (OPR) is defined as an interval of p_i in which a definite type of search plan is optimal, and an OPR is named after the type of its optimal search plan. The sequence of OPRs obtained by varying p_i from 0 to 1 is called the structure of the OPR. The structure of the OPR in the preceding example is $[S_2^0, S_1^0]$. The following is a fundamental theorem for the structure of an OPR.

<u>Theorem 1.14.</u> 1. *If the following conditions are satisfied, an S_i OPR exists.*

(1.22) $$R_i - \frac{c_i}{\alpha_i} > 0,$$

(1.23) $$\frac{c_i}{\alpha_i}\log\left[\frac{\alpha_j}{c_j}\left(R_i - \frac{c_i}{\alpha_i}\right)\right] \leqq \frac{c_1}{\alpha_1} + \frac{c_2}{\alpha_2} - R_j,$$

or

(1.22) $$R_i - \frac{c_i}{\alpha_i} > 0,$$

(1.23a) $$\frac{c_i}{\alpha_i}\log\left[\frac{\alpha_j}{c_j}\left(R_i - \frac{c_i}{\alpha_i}\right)\right] > \frac{c_1}{\alpha_1} + \frac{c_2}{\alpha_2} - R_j,$$

(1.24) $$R_j\left[R_i - \frac{c_i}{\alpha_i}\right] \leqq R_i\left(\frac{c_j}{\alpha_j} - \frac{c_i}{\alpha_i}\right) + \frac{c_i}{\alpha_i}\left[\frac{c_1}{\alpha_1} + \frac{c_2}{\alpha_2} + \frac{c_j}{\alpha_j}\log\left\{\frac{\alpha_j}{c_j}\left(R_i - \frac{c_i}{\alpha_i}\right)\right\}\right].$$

2. *If the inequalities in Eqs.(1.22) and (1.23a) are satisfied, there exists an S_i^0 OPR in the neighborhood of $p_i \fallingdotseq 1$.*

3. *If and only if $R_i - c_i/\alpha_i \leqq 0$, an S_0 OPR exists in the neighborhood of $p_i \fallingdotseq 1$.*

4. *The lower limit of p_i of an S_i OPR is $p_i(T_i{}^e)$, and that of an $S_i{}^0$ is $p_i(\infty)$.*

Proof. 1. If the inequality in Eq.(1.22) is satisfied, both $f(S_i|p_i)$ and $f(S_i{}^0|p_i)$ are negative at $(p_i=1, p_j=0)$ from Corollary 1.13. Therefore, to stop immediately is not optimal in the neighborhood of $p_i \fallingdotseq 1$. The inequality of Eq.(1.23) is derived from $f(S_i|p_i) \leqq f(S_i{}^0|p_i)$. Hence if Eqs.(1.22) and (1.23) are satisfied, there exists an S_i OPR, and if the inequality sign is reversed in Eq.(1.23), $f(S_i|p_i) > f(S_i{}^0|p_i)$ and then $S_i{}^0$ is optimal in the neighborhood of $p_i \fallingdotseq 1$. As mentioned before, the curve $f(S_i{}^0|p_i)$ connects smoothly with $f(S_j{}^0|p_i)$ at $p_i(T_i{}^0)$ given by Corollary 1.12-3. On the other hand, as shown in Fig. 1.1, if $f(S_j{}^0|p_i(T_i{}^e)) > 0$, there exists an interval of p_i in which $f(S_j{}^0|p_i) > f(S_i|p_i)$, and in this case the structure of the OPR has a form $[\ldots, S_0, S_i, S_j{}^0, S_i{}^0]$. (Fig. 1.1 shows the case $i = 1$.) The inequality in Eq.(1.24) is obtained from $f(S_j{}^0|p_i(T_i{}^e)) > 0$.

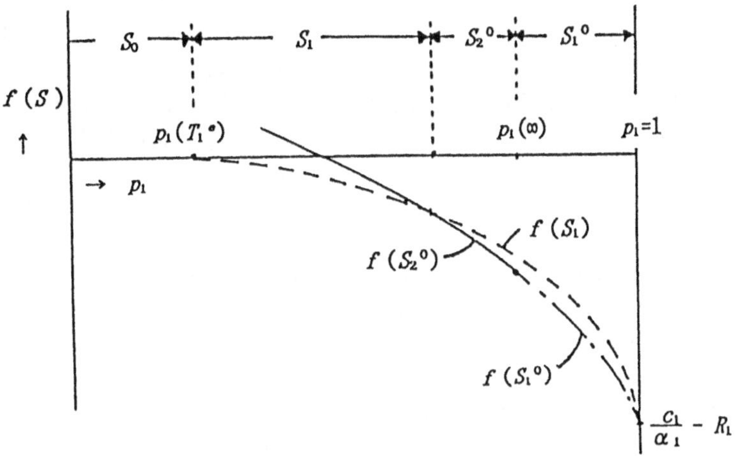

Figure 1.1. $f(S|p_1)$.

2. If the inequalities in Eqs.(1.22) and (1.23a) are satisfied,

$$0 = f(S_0|p_i) > f(S_i|p_i) > f(S_i{}^0|p_i),$$

at $p_i \fallingdotseq 1$, as mentioned earlier. Therefore, $S_i{}^0$ is optimal in the neighborhood of $p_i \fallingdotseq 1$.

3. If $R_i - c_i/\alpha_i \leqq 0$, to stop immediately is obviously optimal, at least in the neighborhood of $p_i \fallingdotseq 1$. On the other hand, if S_0 is optimal at $p_i = 1$, then $f(S_i|p_i=1) = f(S_i{}^0|p_i=1) = c_i/\alpha_i - R_i \geqq f(S_0|p_i) = 0$.

4. Suppose an initial state p_i belongs to an S_i OPR. In this case, if the search is continued without detection, the state variable $p_i(t)$

decreases continuously from p_i to $p_i(T_i^e)$. Therefore, the state in the interval $[p_i(T_i^e),\ p_i]$ also belongs to an S_i OPR. Since $f(S_i|p_i) = f(S_0|p_i) = 0$, if $p_i \leqq p_i(T_i^e)$ from Corollary 1.12, $p_i(T_i^e)$ is the infimum of the S_i OPR.

A similar pattern of reasoning can be applied to prove the infimum $p_i(\infty)$ of an S_i^0 OPR. □

Without any loss of generality, we assume that c_1/α_1 is not smaller than c_2/α_2, and the state variable is p_1. We define the inequalities in Eqs.(1.22a),(1.23a), and (1.24a) as the relations corresponding to Eqs.(1.22),(1.23), and (1.24) with opposite inequality signs, respectively.

From Corollary 1.13 and Theorem 1.14, the conditions specifying the structure of OPR are derived by considering combinations of the inequalities in Eqs.(1.22),(1.23),(1.24),(1.22a),(1.23a), and (1.24a) for $i = 1,2$. The structure of the OPR is classified in eight types and the conditions of the system parameters corresponding to each type are obtained as shown in Table 1.1. In Table 1.1, the notation $[S_2^0,\ S_1^0]$, for example, implies that plan S_2^0 is optimal in the interval $0 \leqq p_1 \leqq p_1(T_i^0)$ and S_1^0 in $p_1(T_i^0) < p_1 \leqq 1$. The expected risk corresponding to S_0, S_i, and S_i^0 are abridged as f_0, f_i, and f_i^0, respectively. Number of Eq. is also abbreviated to the last number, for example the numbers of Eq.(1.22) and (1.22a) are simply denoted by (22) and (22a), respectively.

Fig. 1.2 visualizes the conditions shown in Table 1.1 by use of the R_1-R_2 plane. In Fig. 1.2, the notation, [22(1)], for example, means that the specified curve is given by Eq.(1.22) with $i = 1$ and its inequality sign is replaced by the equality sign. The numbers I - VIII correspond to the type in Table 1.1. As shown in Fig. 1.2, the R_1-R_2 plane is covered exhaustively by disjoint regions each of which corresponds to a type of OPR structure. This implies that all structure of OPR are listed exhaustively with the conditions given in Table 1.1. Thus, the optimal search plan and the associated expected risk are completely determined explicitly as given in Table 1.1 for any set of search parameters, α_i, c_i, p_i and R_i.

Table 1.1. The structure of the optimal plan region

Type NO.	Condition	OPR structure	f versus p_l curve
I	Eqs. (22) and (23a) for both i=1 and 2.	$[S_2^0, S_1^0]$	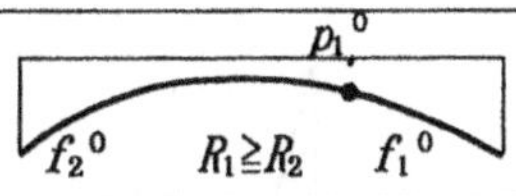 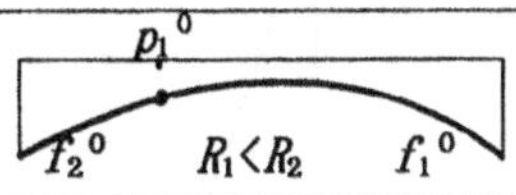
II	Eq. (23) for i=2, Eqs. (23a) and (24a) for i=1, and Eqs. (22) for both i=1, 2.	$[S_2, S_0, S_2^0, S_1^0]$	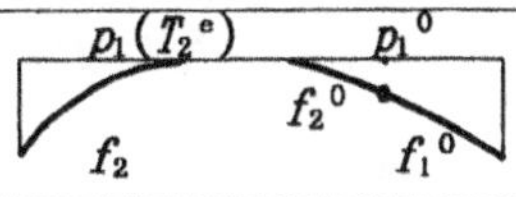
	Same, with box numbers 1 and 2 interchanged.	$[S_2^0, S_1^0, S_0, S_1]$	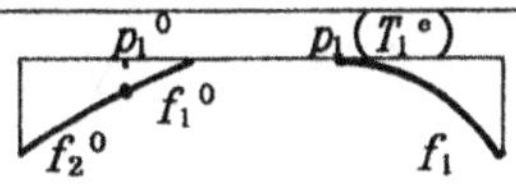
III	Eq. (23) for i=2, Eqs. (23a) and (24) for i=1, and Eqs. (22) for both i=1, 2.	$[S_2, S_0, S_1, S_2^0, S_1^0]$	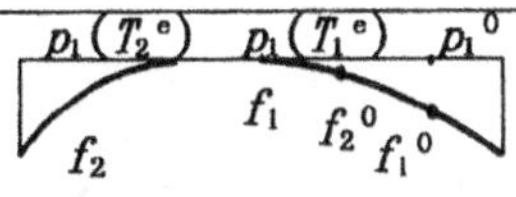
	Same, with box numbers 1 and 2 interchanged.	$[S_2^0, S_1^0, S_2, S_0, S_1]$	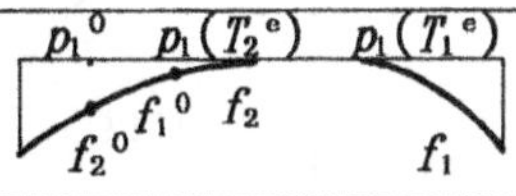
IV	Eqs. (22), (23a), and (24a) for i=1, and Eq. (22a) for i=2.	$[S_0, S_2^0, S_1^0]$	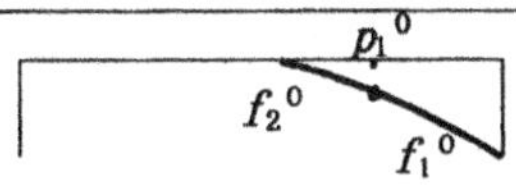
	Same, with box numbers 1 and 2 interchanged.	$[S_2^0, S_1^0, S_0]$	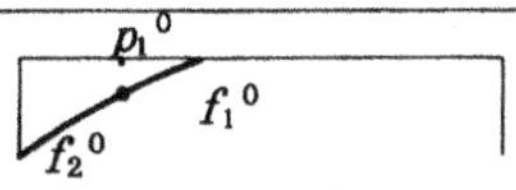
V	Eqs. (22), (23a), and (24) for i=1, Eq. (22a) for i=2.	$[S_0, S_1, S_2^0, S_1^0]$	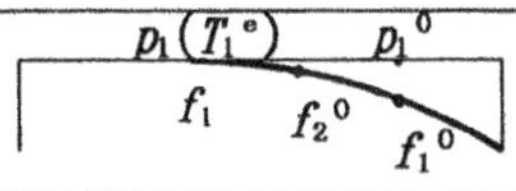
	Same, with box numbers 1 and 2 interchanged.	$[S_2^0, S_1^0, S_2, S_0]$	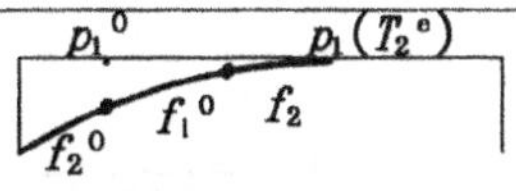
VI	Eqs. (22) for both i=1, 2, and Eqs. (23) for both i=1, 2.	$[S_2, S_0, S_1]$	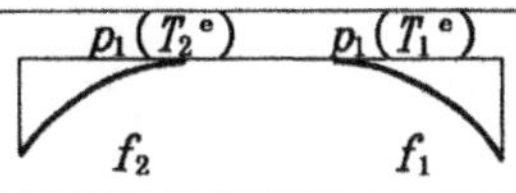
VII	Eq. (22a) for i=2, Eqs. (22) and (23) for i=1.	$[S_0, S_1]$	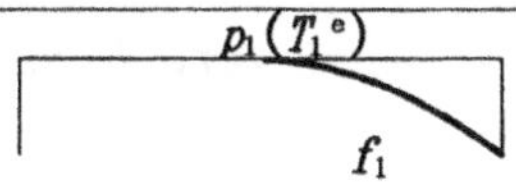
	Same, with box numbers 1 and 2 interchanged.	$[S_2, S_0]$	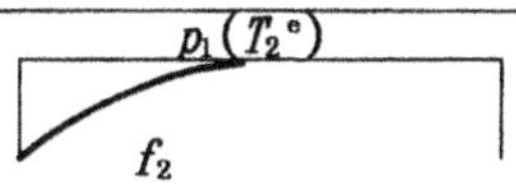
VIII	Eqs. (22a) for both i=1, 2.	$[S_0]$	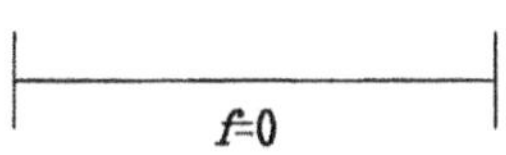

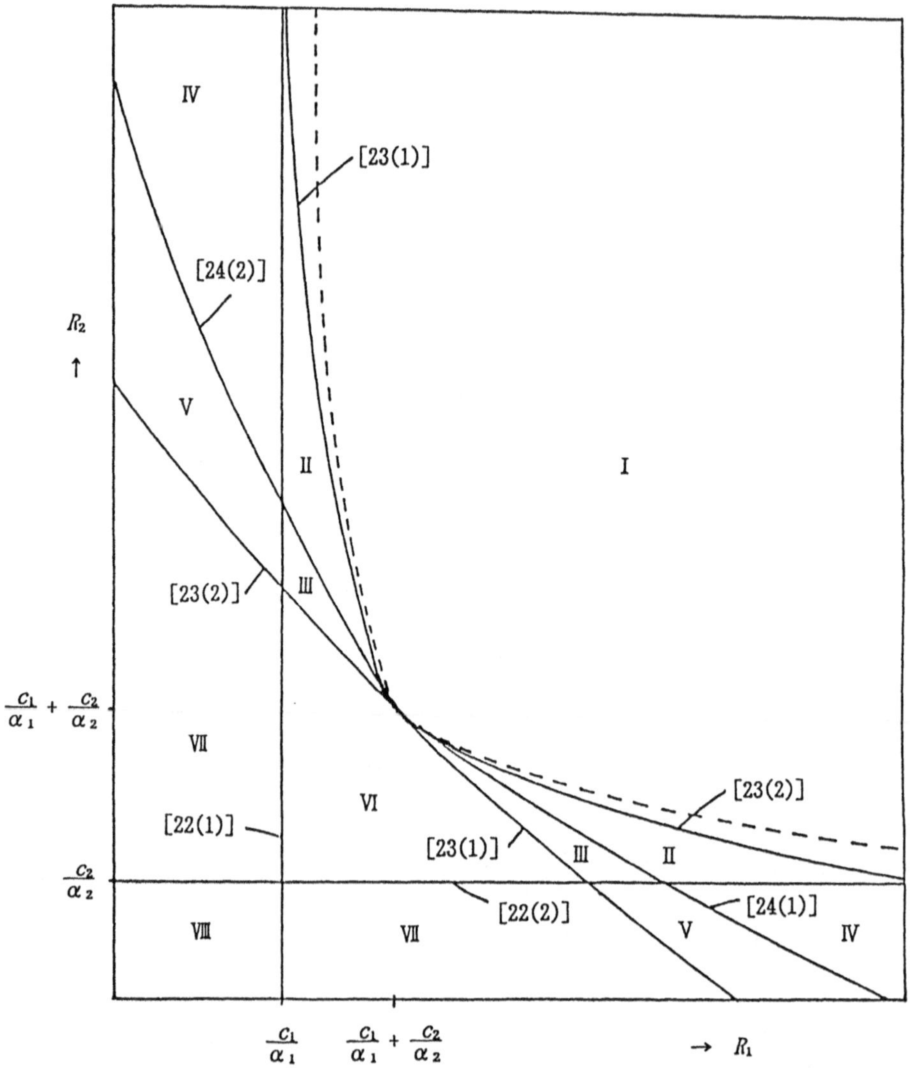

Figure 1.2. The optimal plan region in R_1-R_2 plane.

CHAPTER 2

Optimal Search Plan for a Moving Target

2.1. Introduction

In this chapter, we investigate an optimal search plan for a moving target that maximizes the detection probability of the target or minimizes the expected risk.

From the beginning of the 1970s, studies of the optimal search for a moving target made rapid progress. Pollock (1970) investigated a search problem for a target that moves between two boxes in Markovian fashion. He gave the optimal search plan that maximizes the detection probability with a given number of looks and the optimal search plan that minimizes the expected number of looks. Later, Dobbie (1974) dealt with a time-continuous version of Pollock's two-box model. Hellman (1970, 1971, 1972) studied an optimal search for a target that moves according to a diffusion process in the continuous target space and found a necessary condition for the optimal search plan that maximizes the detection probability. Iida (1972) derived necessary and sufficient conditions for the optimal distribution of continuous searching effort for a target moving along a path selected among a given set of paths with a known probability. Saretsalo (1973) dealt with a search model in which the target motion belongs to a large class of Markov process. Nakai (1973) considered a three-box model under an assumption of perfect detection for the conditional detection function. Stone and Richardson (1974) investigated a search problem for a special class of moving target called the conditionally deterministic motion target, and thereafter, Stone (1977) and Purshiheimo (1977) extended the theory to more general ones. Kan (1977) studied an n-box version of Pollock's model and obtained theorems for the optimal search plan that maximizes the detection probability or minimizes the expected searching cost for a target with Markovian motion. He also discussed the optimal stopping of search considering the searching cost and the reward earned at target detection. Nakai (1980) investigated an optimal search plan with a wait in searching for a moving target. Brown (1980) studied an optimal search with continuous searching effort and proposed an efficient algorithm for calculating the optimal distribution of searching effort.

The algorithm is essentially the same as that employed by Iida (1972) in computing his numerical example and was later generalized and named the FAB algorithm by Washburn (1983). Washburn (1980a) dealt with a discrete search problem in which only one box is looked at each time until T. Washburn (1981b) also gave an upper bound on the detection probability for this model and proposed a generalized FAB algorithm (1983). A generalized theorem for the maximization problem of the detection probability for a moving target was obtained by Stone (1979). He derived necessary and sufficient conditions for the optimal distribution of searching effort assuming a regular detection function and weak conditions on the stochastic process used to model the target motion, by which practically no restrictions are imposed in real-world applications. Stromquist and Stone (1981) generalized the theory to include a wide class of nonlinear and nonseparable functional and separable constraints. Their theorem can be applied not only to the maximization problem of the detection probability for a moving target, but also to a large variety of optimal search problems, such as minimizing the expected risk or the like. A whereabouts search model for a moving target was investigated by Stone and Kadane (1981). Tierney and Kadane (1983) investigated a generalized surveillance search model for a target with Markovian motion, which includes the detection search model and the whereabouts search model as special cases. They proposed an improved FAB algorithm in calculating the optimal search plan. Eagle (1984) also studied an optimal search for a moving target. In his model, he introduces a restriction on the searcher such that the box searched at a given time point should be within some neighborhood of the box searched at the previous time point. Iida and Hozaki (1988) obtained the optimal search plan minimizing the expected risk and Iida (1989b) derived a closed form solution of the optimal search plan for the conditionally deterministic motion target.

In the next section, we will derive the optimal search plan for a moving target that maximizes the detection probability with a definite amount of continuous searching effort. Then, in Section 2.3, the optimal search plan that minimizes the expected risk will be investigated. The final section in this chapter deals with a search problem for a target with conditionally deterministic motion.

2.2. Optimal Search Plan Maximizing the Detection Probability

The problem investigated in this section, which is taken from Iida (1972), is as follows.

The target space is assumed to consist of n boxes. The time space is discrete and the search is conducted at each time point $t = 1, 2, \cdots, T$.

An appeared target moves in the target space, selecting a path Z_k from a finite set of paths $\mathbf{Z} = \{Z_k, k=1, 2, \cdots, m\}$ with a probability p_k (>0), $\Sigma_k p_k = 1$. The path Z_k is defined as a sequence of boxes containing the target at each t: $Z_k = \{Z_k^t, t=1, 2, \cdots, T\}$, where Z_k^t is the box in which the target exists at time point t if the path Z_k is selected. We assume that $\mathbf{Z}$ and $\{p_k\}$ are known to the searcher. A searching effort $\Phi(t)$ is available to the searcher at t and $\{\Phi(t), t=1, 2, \cdots, T\}$ is given to the searcher prior to the beginning of the search. $\Phi(t)$ is assumed to be divisible arbitrarily in distribution to boxes. Let $\phi(i, t)$ be searching effort density allocated to box i at time point t and let A_i be the area of box i. A set $\boldsymbol{\phi} = \{\phi(i, t), i=1, 2, \cdots, n, t=1, 2, \cdots, T\}$ is called a search plan. A constraint on the total searching effort, $\Sigma_i A_i \phi(i, t) \leq \Phi(t)$, is imposed at each t. The conditional detection function given that the target is really in box i at t and $\phi(i, t)$ is allocated to box i at t is denoted by $b(i, \phi(i, t))$. The function $b(i, \cdot)$ is assumed to be regular. Here we assume that the detection probability of the target during the search operation is adopted as the measure of effectiveness and the searcher wishes to maximize it.

The modeling of the search process is as follows. At the beginning of the search, the target selects a path Z_k with probability p_k. The position of the target at $t = 1$ is box Z_k^1. The searcher decides allocation of his searching effort $\Phi(1)$ and the search at $t = 1$ is conducted with the effort distribution $\{A_i \phi(i, 1)\}$. If the target is detected, the search is terminated successfully. Otherwise, the search process enters the next time point $t = 2$, and the target moves to box Z_k^2, and then the search with effort distribution $\{A_i \phi(i, 2)\}$ is conducted. This move-and-search sequence is continued until the detection of the target or the end of the search, $t = T$, whichever comes first.

The detection probability $P(\boldsymbol{\phi})$ of the target until T is given by

(2.1) $$P(\boldsymbol{\phi}) = \Sigma_k p_k\{1-\Pi_t(1-b(Z_k^t, \phi(Z_k^t, t)))\}.$$

The searching effort density $\{\phi(i, t)\}$ is constrained by inequalities,

(2.2a) $$\Sigma_i A_i \phi(i, t) \leq \Phi(t) \quad \text{for all } t = 1, 2, \cdots, T,$$

(2.2b) $$\phi(i, t) \geq 0 \quad \text{for all } i \text{ and } t.$$

Therefore, our problem is formulated as a nonlinear programming problem to obtain the searching effort distribution $\{\phi^*(i, t)\}$ that maximizes the detection probability $P(\boldsymbol{\phi})$ given by Eq.(2.1) subject to the constraints in Eqs.(2.2a) and (2.2b). For this problem, we have the following theorem.

<u>Theorem 2.1.</u> *Let* $\mathbf{I}$ *be a set of paths* $\{Z_k | Z_k^t = i\}$ *and* $b'(i, \phi(i, t))$ *be the derivative of* $b(i, \cdot)$ *with respect to* $\phi(i, t)$. *A necessary and sufficient condition for the optimal distribution of searching effort is,*

if $\phi^*(i,t) > 0$,

(2.3a) $$\frac{b'(i,\phi^*(i,t))}{A_i\{1-b(i,\phi^*(i,t))\}} \Sigma_{Z_k\in I}\, p_k \prod_{j=1}^{T} \{1-b(Z_k{}^j,\phi^*(Z_k{}^j,j))\} = \lambda(t),$$

if $\phi^*(i,t) = 0$,

(2.3b) $$\frac{b'(i,0)}{A_i} \Sigma_{Z_k\in I}\, p_k \prod_{j=1}^{T} \{1-b(Z_k{}^j,\phi^*(Z_k{}^j,j))\} \leqq \lambda(t)$$

for all i and t, and

(2.4) $$\Sigma_i A_i \phi^*(i,t) = \Phi(t).$$

$\lambda(t)$ *is nonnegative and is determined by Eq.(2.4).*

Proof. The theorem is directly obtained by applying the Kuhn-Tucker theorem. Since the constraints in Eqs.(2.2a) and (2.2b) are linear and the objective function given by Eq.(2.1) is concave with respect to $\phi(i,t)$ by the assumption of regularity of the detection function, the conditions in Eqs.(2.3a),(2.3b), and (2.4), are the necessary and sufficient conditions for the optimal search plan $\{\phi^*(i,t)\}$ by the Kuhn-Tucker theorem. □

Remark 2.1. Theorem 2.1 is interpreted as follows. Consider a target selecting a path $Z_k \in \mathbf{I}$. The probability that the target is detected only at time point t in box i during the search is obtained by

$$P(i,t,Z_k) = b(i,\phi^*(i,t))\prod_{j\neq t}\{1-b(Z_k{}^j,\phi^*(Z_k{}^j,j))\}.$$

Differentiating $P(i,t,Z_k)$ with respect to $\phi^*(i,t)$, we have

(2.5) $$\frac{\partial P(i,t,Z_k)}{A_i\partial\phi^*(i,t)} = \frac{b'(i,\phi^*(i,t))\prod_j\{1-b(Z_k{}^j,\phi^*(Z_k{}^j,j))\}}{A_i\{1-b(i,\phi^*(i,t))\}}.$$

This quantity is the increment of the detection probability by unit increase of the searching effort density at t in i, namely, the marginal detection probability conditional on the target selecting $Z_k \in \mathbf{I}$. Using this relation, we can rewrite Eqs.(2.3a) and (2.3b) as follows: if $\phi^*(i,t) >(=) 0$,

$$\Sigma_{Z_k\in I}\, p_k \frac{\partial P(i,t,Z_k)}{A_i\partial\phi^*(i,t)} =(\leq)\ \lambda(t).$$

Hence, the meaning of Theorem 2.1 is described as follows. If box i should be searched at t, the density of searching effort should be determined in such a way that the expected marginal detection probability is balanced to $\lambda(t)$ among the boxes being searched at t. If box i should not be searched at t, the reason is that box i does not have a larger expected marginal detection probability than $\lambda(t)$ at that time.

The optimal search plan determined by Theorem 2.1 is not the myopic search plan. This is easily confirmed by noting that the left-hand side of Eqs.(2.3a) and (2.3b) contain the search plan $\{\phi(i,j)\}$ in future time points $j = t+1,\cdots,T$.

Remark 2.2. In this section, we consider the problem with the discrete target space and the discrete time space. However, these spaces can be modified easily to the continuous spaces and we can derive similar conditions to those in Eqs. (2.3a), (2.3b) and (2.4). In this case, the problem is formulated as a variational problem and the optimal condition is derived by considering a feasible variation similar to the derivation of Theorem 1.1.

2.3. Optimal Search Plan Minimizing the Expected Risk

2.3.1. Optimal Distribution of Searching Effort

In this section, we will study the optimal distribution of searching effort that minimizes the expected risk of search for an appeared moving target. The problem dealt with in this section, which is summarized from Iida and Hozaki (1988), is described in detail as follows.

1. We consider a target space $\mathbf{Y}$ with σ-finite measure ξ and a continuous time interval $\mathbf{T} = [0, T]$ of search time. Let μ be the product measure on $\mathbf{Y} \times \mathbf{T}$. Let $\mathbf{Z}$ be a μ-measurable subset of $\mathbf{Y} \times \mathbf{T}$ and denote its t section by $\mathbf{Z}_t = \{y \in \mathbf{Y} \mid (y, t) \in \mathbf{Z}\}$.

2. Suppose the target's motion is given by a stochastic process $\mathbf{X} = \{X_t \in \mathbf{Z}\}$, where X_t is the position of the target at t. We assume that X_t has a probability density function $q_t(X_t = y)$ defined on $\mathbf{Z}_t$.

3. We assume that the search is started at t_0 and the searcher may stop the search at any time when he thinks the search does not pay. C: $\mathbf{T} \to (0, \infty)$ is assumed to be given, where $C(t)$ is total searching cost rate available to the searcher at t and is assumed to be continuously divisible in the target space.

4. A search plan is a Borel measurable function $\phi : \mathbf{Z} \to [0, \infty)$ that specifies an allocation of searching effort density in both time and space.

5. Let $\alpha : \mathbf{Z} \to (0, \infty)$ be μ-measurable. For each sample path of the process X, the detection probability of the target given that the target follows the path is a function of the weighted total searching effort density ψ,

$$\psi = \int_{t_0}^{t} \alpha(X_s, s)\, \phi(X_s, s)\, ds,$$

which is the cumulative searching effort density on the target over the path until t. The weight $\alpha(y, t)$ represents the relative detectability of the target being at (y, t). Let b: $\psi \to [0, 1]$ be the probability of detecting the target up to t given that the target follows the path and effort density ψ is applied. The detection function $b(\psi)$ is assumed to be regular.

6. R: $\mathbf{T}\to[0,\infty)$ is assumed to be given. $R(t)$ is a reward earned by the searcher if the target is detected at t. $R(t)$ is assumed to be a nonnegative, nonincreasing, and differentiable function of t.

7. Let c_0: $\mathbf{Z}\to(0,\infty)$ be μ-measurable. $c_0(y, t)$ is the unit cost of searching effort density at $(y, t)\in\mathbf{Z}_t$.

8. The expected risk of the search is adopted as the measure of effectiveness for a search plan. The search plan that gives the minimum expected risk is called optimum.

Under these assumptions, conditions for the optimal search plan are derived in the following two steps. First we deal with a problem conditional on a finite T (T is called the limit time of the search) and the conditionally optimal distribution $\boldsymbol{\phi}_T^*$ of searching effort is sought. Then, we consider a problem with no restriction on the search time and determine the optimal stopping time T^*. The search plan $\boldsymbol{\phi}_{T*}^*$ is the (unconditionally) optimal search plan.

Suppose a limit time T is given and a search plan $\boldsymbol{\phi}_T=\{\phi_T(y,t),(y,t)\in\mathbf{Z}\}$ is employed. The search plan $\boldsymbol{\phi}_T$ must satisfy the following constraints from the third assumption.

(2.6)
$$\int_{Z_t} c_0(y,t)\phi_T(y,t)\,d\xi(y)\le C(t),$$
$$\phi_T(y,t)\ge 0,$$

for a.e. $t\in\mathbf{T}$ and all $(y,t)\in\mathbf{Z}$. The search plan $\boldsymbol{\phi}_T$, which satisfies the conditions for all t is called a feasible plan and the set of the feasible plan is denoted by $\boldsymbol{\Psi}$. $\boldsymbol{\Psi}_0$ is defined as the set of $\boldsymbol{\phi}_T\in\boldsymbol{\Psi}$ such that

(2.7)
$$\int_{Z_t} c_0(y,t)\phi_T(y,t)\,d\xi(y)=C(t).$$

From now on, our attention is concentrated on the plan $\boldsymbol{\phi}_T\in\boldsymbol{\Psi}$ or $\boldsymbol{\Psi}_0$. The detection probability $P_t(\boldsymbol{\phi}_T)$ until t is given by

$$P_t(\boldsymbol{\phi}_T)=E[b(\int_{t0}^{t}\alpha(X_s,s)\phi_T(X_s,s)\,ds)],$$

where $E[\cdot]$ is the expectation over the sample paths of $\mathbf{X}$.

The cumulative searching cost $L(t,\boldsymbol{\phi}_T)$ up to t and its derivative with respect to t are given by

$$L(t,\boldsymbol{\phi}_T)=\int_{t0}^{t}\int_{Z_s} c_0(y,s)\phi_T(y,s)\,d\xi(y)\,ds,$$
$$L'(t,\boldsymbol{\phi}_T)=\frac{\partial L(t,\boldsymbol{\phi}_T)}{\partial t}=\int_{Z_t} c_0(y,t)\phi_T(y,t)\,d\xi(y).$$

Using $P_t(\boldsymbol{\phi}_T)$ and $L(t,\boldsymbol{\phi}_T)$, we obtain the expected risk $f(\boldsymbol{\phi}_T)$ of the search plan $\boldsymbol{\phi}_T$ as

$$f(\boldsymbol{\phi}_T)=\int_{s:[t0,T]}(L(s,\boldsymbol{\phi}_T)-R(s))\,dP_s(\boldsymbol{\phi}_T)+L(T,\boldsymbol{\phi}_T)(1-P_T(\boldsymbol{\phi}_T)).$$

Since $b(\cdot)$ is regular, $P_t(\boldsymbol{\phi}_T)$ is continuous with respect to t. Integration by parts yields

(2.8) $$f(\boldsymbol{\phi}_T) = \int_{t0}^{T} (R'(s)-L'(s, \boldsymbol{\phi}_T))P_s(\boldsymbol{\phi}_T)\,ds + L(T, \boldsymbol{\phi}_T) - R(T)P_T(\boldsymbol{\phi}_T).$$

Hence, the problem with a finite limit time T is formulated as a variational problem to find the function $\boldsymbol{\phi}_T^*$ that minimizes the functional $f(\boldsymbol{\phi}_T)$ given by Eq.(2.8) subject to the constraints in Eq.(2.6).

$$f(\boldsymbol{\phi}_T^*) = \inf_{\phi_T} f(\boldsymbol{\phi}_T) \quad \text{for } \boldsymbol{\phi}_T \in \boldsymbol{\Psi}.$$

$\boldsymbol{\phi}_T^*$ is called the R_T-optimal allocation of searching effort. In the second stage of our optimization, the limit time T is considered as a variable in $[t_0, \infty)$. The earliest T that minimizes $f(\boldsymbol{\phi}_T^*)$,

$$f(\boldsymbol{\phi}_{T*}^*) = \inf_{T \geq t0} \inf_{\phi_T} f(\boldsymbol{\phi}_T)$$

is defined as the optimal stopping time T^* of the search.

(2.9) $$T^* = \min\{\, T \mid \inf_T f(\boldsymbol{\phi}_T^*)\}.$$

From now on, we omit the subscript T in $\phi_T(x, t)$ and $\boldsymbol{\phi}_T$, if any confusion is not expected.

Before presenting a theorem, we define the Gateaux differential $\delta g(\boldsymbol{\phi}, e)$ of a real-valued functional $g(\boldsymbol{\phi})$ at $\boldsymbol{\phi}$ in a direction e by

(2.10) $$\delta g(\boldsymbol{\phi}, e) = \lim_{\varepsilon \to 0} \left\{ \frac{g(\boldsymbol{\phi}+\varepsilon e)-g(\boldsymbol{\phi})}{\varepsilon} \right\}$$

for $\boldsymbol{\phi}$ and $\boldsymbol{\phi}+\varepsilon e \in \boldsymbol{\Psi}$ for all sufficiently small and positive ε. Suppose that there exists a function $d(\boldsymbol{\phi}, y, t)$ defined on $\mathbf{Z}$ such that the Gateaux differential is given by

(2.11) $$\delta g(\boldsymbol{\phi}, e) = \int_Z d(\boldsymbol{\phi}, y, t)e(y, t)\,d\mu(y, t)$$

for every e and $\boldsymbol{\phi} \in \boldsymbol{\Psi}$. Then $d(\boldsymbol{\phi}, y, t)$ is called a kernel of the Gateaux differential at $\boldsymbol{\phi}$.

Very general treatments of the moving-target search are provided by Stromquist and Stone (1981) allowing non-Markovian motion and non-exponential detection function. They dealt with a maximization problem for a real-valued functional $F(\boldsymbol{\phi})$ under the constraints in Eq.(2.6) and derive necessary and sufficient conditions for optimum. The necessary and sufficient conditions for $\boldsymbol{\phi}^*$ to be optimal given by theorems 1 and 2 in Stromquist and Stone (1981) are quoted as the next lemma.

<u>Lemma 2.2.</u> *Let $\boldsymbol{\Psi}$ be the set of measurable functions $\phi : \mathbf{Z} \to [0, \infty)$ satisfying Eq.(2.6), let $\boldsymbol{\phi}^* \in \boldsymbol{\Psi}$, and let F be a real-valued functional on $\boldsymbol{\Psi}$. Assume that F has a Gateaux differential at $\boldsymbol{\phi}^*$ with kernel $d(\boldsymbol{\phi}^*, y, t)$. Then a necessary condition for $\boldsymbol{\phi}^*$ to be optimal is that there exists a measurable function $\lambda : \mathbf{T} \to (-\infty, \infty)$ such that for a.e. (y, t)*

$$d(\boldsymbol{\phi}^*, y, t) \leq \lambda(t) c_0(y, t) \quad \text{if } \phi^*(y, t) = 0,$$
$$d(\boldsymbol{\phi}^*, y, t) = \lambda(t) c_0(y, t) \quad \text{if } \phi^*(y, t) > 0.$$

In addition to the hypotheses of the necessary condition mentioned earlier, assume that $F(\boldsymbol{\phi})$ *is concave and* $\boldsymbol{\phi}^* \in \boldsymbol{\Psi}_0$ *for all* t. *Then the necessary conditions are also sufficient for* $\boldsymbol{\phi}^*$ *to be optimal.*

Letting $F(\boldsymbol{\phi}) = -f(\boldsymbol{\phi})$ given by Eq.(2.8), we can apply Lemma 2.2 to our problem. The following theorem gives necessary and sufficient conditions for $\boldsymbol{\phi}^*$ to be conditionally optimal.

Theorem 2.3. *A necessary condition for the* R_T*-optimal search plan* $\boldsymbol{\phi}_T^*$ *is that there exists a nonnegative function* $\lambda(t)$, $\{\lambda(t), t \in \mathbf{T}\} \neq \{0\}$ *such that for a.e.* y *and* t,

(2.12a) $\quad$ if $\phi_T^*(y, t) > 0 \quad -A_{yt}{}^T(\boldsymbol{\phi}_T^*) = \lambda(t) c_0(y, t)$,

(2.12b) $\quad$ if $\phi_T^*(y, t) = 0 \quad -A_{yt}{}^T(\boldsymbol{\phi}_T^*) \leq \lambda(t) c_0(y, t)$,

where

$$A_{yt}{}^T(\boldsymbol{\phi}) = \int_t^T (R'(s) - L'(s, \boldsymbol{\phi})) D_{yt}{}^s(\boldsymbol{\phi}) ds$$

(2.13)
$$+ c_0(y, t)(1 - P_t(\boldsymbol{\phi})) - R(T) D_{yt}{}^T(\boldsymbol{\phi}),$$

$$D_{yt}{}^T(\boldsymbol{\phi}) = \alpha(y, t) E_{yt}[b'(\int_{t0}^T \alpha(X_s, s) \phi(X_s, s) ds)] q_t(y),$$

and $E_{yt}[\cdot]$ *denotes the expectation over the sample paths of* $\mathbf{X}$ *conditional on* $X_t = y$.

If $\lambda(t) > 0$, $t \in \mathbf{T}$, *then the search plan obtained from Eq.(2.12) is an element of* $\boldsymbol{\Psi}_0$.

If $f(\boldsymbol{\phi})$ *is convex and* $\boldsymbol{\phi}^* \in \boldsymbol{\Psi}_0$ *in* $\mathbf{T}$, *the necessary conditions are also sufficient for* $\boldsymbol{\phi}^*$ *to be conditionally optimal.*

Proof. The kernel of the Gateaux differential of the functional $f(\boldsymbol{\phi})$ at $\boldsymbol{\phi}$ defined by Eqs.(2.10) and (2.11) is calculated as $A_{yt}{}^T(\boldsymbol{\phi})$ given by Eq.(2.13). To derive it, we calculate the Gateaux differential of each term in Eq.(2.8). Since the detection function $b(\psi)$ is regular by the assumption, $b(\psi)$ is differentiable and has a bounded positive derivative $b'(\psi)$. Hence the Gateaux differential of $P_t(\boldsymbol{\phi})$ exists and is obtained as

$$\delta P_t(\boldsymbol{\phi}, e) = \int_{t0}^t \int_{X_s} \alpha(y, s) E_{ys}[b'(\int_{t0}^t \alpha(X_z, z) \phi(X_z, z) dz)]$$
$$\times \; q_s(y) e(y, s) d\xi(y) ds$$
$$= \int_{t0}^t \int_{X_s} D_{ys}{}^t(\boldsymbol{\phi}) e(y, s) d\xi(y) ds.$$

Since $L(t, \boldsymbol{\phi})$ and $L'(t, \boldsymbol{\phi})$ are linear with respect to $\phi(y, t)$, we have

$$\delta L(t, \boldsymbol{\phi}, e) = \int_{t0}^t \int_{X_s} c_0(y, s) e(y, s) d\xi(y) ds,$$
$$\delta L'(t, \boldsymbol{\phi}, e) = \int_{X_t} c_0(y, t) e(y, t) d\xi(y).$$

Therefore, the Gateaux differential of $f(\boldsymbol{\phi})$ is obtained as follows.

$$\begin{aligned}\delta f(\boldsymbol{\phi}, e) &= \int_{t0}^{T}\int_{Zs}^{T}[\int s(R'(x)-L'(x, \boldsymbol{\phi}))D_{y,s}{}^{x}(\boldsymbol{\phi})dx \\ &\quad + c_0(y, s)(1-P_s(\boldsymbol{\phi}))-R(T)D_{y,s}{}^{T}(\boldsymbol{\phi})]e(y, s)d\xi(y)ds \\ &= \int_{t0}^{T}\int_{Zs} A_{y,s}{}^{T}(\boldsymbol{\phi})e(y, s)d\xi(y)ds.\end{aligned}$$

If we set $F(\boldsymbol{\phi}) = -f(\boldsymbol{\phi})$, Lemma 2.2 is applicable to our problem since the objective functional $-f(\boldsymbol{\phi})$ has the Gateaux differential, which is a linear functional defined by integration with the kernel function $-A_{y,t}{}^{T}(\boldsymbol{\phi})$ given by Eq.(2.13). Letting $d(\boldsymbol{\phi}^*, y, t) = -A_{y,t}{}^{T}(\boldsymbol{\phi}^*)$ in Lemma 2.2, the relations in Eqs.(2.12a) and (2.12b) are derived. The nonnegativity of $\lambda(t)$ is proved as follows. We assume $\boldsymbol{\phi}^*(y, t) > 0$ and $\lambda(t) < 0$ in the neighborhood of (y_1, t_1). Consider a search plan $\boldsymbol{\phi}^0$ that differs from the optimal $\boldsymbol{\phi}^*$ only in the neighborhood of (y_1, t_1),

$\phi^0(y, t) = \phi^*(y, t) - \varepsilon$ in the Δz neighborhood of (y_1, t_1),
$\phi^0(y, t) = \phi^*(y, t)$ otherwise.

Then it is easily proved that $\{\phi^0(y, t)\}$ is feasible, and we have

$$f(\boldsymbol{\phi}^0) - f(\boldsymbol{\phi}^*) = -A_{y1t1}{}^{T}(\boldsymbol{\phi}^*)\varepsilon\Delta z$$

for sufficiently small ε (> 0). Since $\phi^*(y, t) > 0$, the right-hand side of the preceding equation is denoted as $\lambda(t_1)c_0(y_1, t_1)\varepsilon\Delta z$ and is negative by the assumption. Hence we have $f(\boldsymbol{\phi}^0) < f(\boldsymbol{\phi}^*)$. This result contradicts the optimality of $\boldsymbol{\phi}^*$, and therefore we can conclude $\lambda(t) \geqq 0$.

It is similarly proved that Eq.(2.7) holds if $\lambda(t) > 0$. Here we assume

$$0 < \int_{Zt} c_0(y, t)\phi^*(y, t)d\xi(y) < C(t) \quad \text{for } \lambda(t) > 0.$$

Then there exists (y_1, t_1) such that $\phi^*(y, t) > 0$ in the neighborhood of (y_1, t_1). We consider a search plan $\boldsymbol{\phi}^{**}$:

$\phi^{**}(y, t) = \phi^*(y, t) + \varepsilon$ in the Δz neighborhood of (y_1, t_1),
$\phi^{**}(y, t) = \phi^*(y, t)$ otherwise.

If ε (>0) is sufficiently small, then $\boldsymbol{\phi}^{**}$ is also a member of $\boldsymbol{\Psi}$ and the following is derived.

$$\begin{aligned}f(\boldsymbol{\phi}^{**}) - f(\boldsymbol{\phi}^*) &= A_{y1t1}{}^{T}(\boldsymbol{\phi}^*)\varepsilon\Delta z \\ &= -\lambda(t_1)c_0(y_1, t_1)\varepsilon\Delta z + o(\varepsilon\Delta t) < 0.\end{aligned}$$

This contradicts the optimality of $\boldsymbol{\phi}^*$, and therefore, Eq.(2.7) is established if $\lambda(t) > 0$.

As stated in Lemma 2.2, the necessary condition is also sufficient if $F(\boldsymbol{\phi}^*)$ is concave (in the maximization problem) and $\boldsymbol{\phi}^* \in \boldsymbol{\Psi}_0$ for all t. However in this statement, the concavity of the objective functional must be changed to the convexity of the expected risk function $f(\boldsymbol{\phi}^*)$ in our minimization problem. □

The following lemma gives a sufficient condition for $f(\boldsymbol{\phi})$ to be convex.

Lemma 2.4. *The function $f(\boldsymbol{\phi})$ given by Eq.(2.8) is convex, if*

$$\int_{t_0}^{T}\{L'(s,\boldsymbol{\phi}^1)-L'(s,\boldsymbol{\phi}^2)\}\{P_s(\boldsymbol{\phi}^2)-P_s(\boldsymbol{\phi}^1)\}ds \geqq 0$$

for any $\boldsymbol{\phi}^1$ and $\boldsymbol{\phi}^2 \in \boldsymbol{\Psi}$.

Proof. Let $\boldsymbol{\phi} = (1-\theta)\boldsymbol{\phi}^1+\theta\boldsymbol{\phi}^2$, $0 \leqq \theta \leqq 1$, where both $\boldsymbol{\phi}^1$ and $\boldsymbol{\phi}^2 \in \boldsymbol{\Psi}$. Then $\boldsymbol{\phi} \in \boldsymbol{\Psi}$ is confirmed by

$$\begin{aligned}\int_{Z_t} c_0(y,t)\phi(y,t)d\xi(y) &= (1-\theta)\int_{Z_t} c_0(y,t)\phi^1(y,t)d\xi(y)\\ &\quad+\theta\int_{Z_t} c_0(y,t)\phi^2(y,t)d\xi(y) \leqq C(t).\end{aligned}$$

By substituting $\boldsymbol{\phi} = (1-\theta)\boldsymbol{\phi}^1+\theta\boldsymbol{\phi}^2$ into Eq.(2.8), we have

$$\begin{aligned}f(\boldsymbol{\phi}) &\leqq (1-\theta)f(\boldsymbol{\phi}^1) + \theta f(\boldsymbol{\phi}^2)\\ &\quad+(1-\theta)\theta\int_{t_0}^{T}(L'(s,\boldsymbol{\phi}^1)-L'(s,\boldsymbol{\phi}^2))(P_s(\boldsymbol{\phi}^1)-P_s(\boldsymbol{\phi}^2))ds,\end{aligned}$$

because of the concavity of $b(\psi)$. Hence if the last term of the right-hand side of the preceding inequality is nonpositive, $f(\boldsymbol{\phi})$ is a convex function of $\boldsymbol{\phi} \in \boldsymbol{\Psi}$. □

Remark 2.3. Let us consider the meaning of Theorem 2.3. According to the theorem, the R_T-optimal distribution of searching effort density is determined by

$$-A_{yt}{}^T(\boldsymbol{\phi}^*) =(\leqq)\ \lambda(t)c_0(y,t), \quad \text{if} \quad \boldsymbol{\phi}^*(y,t) >(=) 0$$

for a.e. $(y,t) \in \mathbf{Z}$. Integrating the right-hand side of $A_{yt}{}^T(\boldsymbol{\phi}^*)$ given by Eq.(2.13) by parts, we obtain

$$\begin{aligned}-A_{yt}{}^T(\boldsymbol{\phi}^*) &= (R(t)-L(t,\boldsymbol{\phi}^*))D_{yt}{}^t(\boldsymbol{\phi}^*) - c_0(y,t)(1-P_t(\boldsymbol{\phi}^*))\\ &\quad+L(T,\boldsymbol{\phi}^*)D_{yt}{}^T(\boldsymbol{\phi}^*)+\int_{S:[t,T]}(R(s)-L(s,\boldsymbol{\phi}^*))dD_{yt}{}^s(\boldsymbol{\phi}^*).\end{aligned}$$

$D_{yt}{}^T(\boldsymbol{\phi}^*)$, the kernel of the Gateaux differential of $P_t(\boldsymbol{\phi}^*)$ given by Eq.(2.13), is the increment of detection probability density when the unit searching effort density is added to $\boldsymbol{\phi}^*$ at (y,t). Hence the first term in the right-hand side of $-A_{yt}{}^T(\boldsymbol{\phi}^*)$, $(R(t)-L(t,\boldsymbol{\phi}^*))D_{yt}{}^t(\boldsymbol{\phi}^*)$, is interpreted as the expected return gained by the additional unit density of searching effort at (y,t). The second term, $-c_0(y,t)(1-P_t(\boldsymbol{\phi}^*))$, is the expected cost of the additional unit searching effort density, and the third term, $L(T,\boldsymbol{\phi}^*)D_{yt}{}^T(\boldsymbol{\phi}^*)$, is the expected search cost in $[t_0,T]$ that is saved by the increment of detection probability density at (y,t). One should note that the sign of this term is positive since the saved cost is considered as profit instead of cost. The last term,

$$\int_{S:[t,T]} (R(s)-L(s,\boldsymbol{\phi}^*))\,dD_{yt}{}^s(\boldsymbol{\phi}^*),$$

represents the variation of the return in the interval $[t,T]$ caused by the additional unit density of searching effort at (y,t). Therefore, $-A_{yt}{}^T(\boldsymbol{\phi}^*)$ is the expected return earned by the searcher if he allocates additional unit searching effort density to $\phi^*(y,t)$ at (y,t). We shall call this the expected marginal return of $\boldsymbol{\phi}^*$ at (y,t). By this reasoning, Theorem 2.3 is interpreted as follows. If searching effort is to be allocated to $(y,t) \in \mathbf{Z}$, the searching effort density should be balanced in such a way that the expected marginal return versus cost ratio, $-A_{yt}{}^T(\boldsymbol{\phi}^*)/c_0(y,t)$, is equal to $\lambda(t)$ in the region being searched at t. If searching effort should not be allocated to (y,t), the point does not have a larger expected marginal return versus cost ratio than $\lambda(t)$.

If $\lambda(t) > 0$, then $-A_{yt}{}^T(\boldsymbol{\phi}^*) = \lambda(t)c_0(y,t)$ in Eq. (2.12a) means that the expected marginal return at t is positive, and therefore, the expected return increases (that is, the expected risk decreases) as the searching effort density increases. Hence, if $\lambda(t) > 0$, the total searching cost rate should be used exhaustively; $\{\phi^*(y,t),(y,t)\in \mathbf{Z}_t\} \in \boldsymbol{\Psi}_0$, if $\lambda(t) > 0$.

Remark 2.4. The R_T-optimal allocation of searching effort given by Theorem 2.3 usually cannot be obtained in analytical form; often we must calculate it numerically. An algorithm for an iterative approximation, called the FAB (forward and backward) algorithm proposed by Brown (1980) and by Washburn (1983) in a more general form, can be applied to calculate the optimal allocation of searching effort given by Theorem 2.3.

Remark 2.5. If the reward earned by the searcher when he detects the target depends on the position y of the target as well as the time t, the problem is considerably complicated. We assume the searcher earns a reward $R(y,t)$ (≥ 0) when the target is detected at (y,t) and other system parameters are identical with the model described earlier. The conditional expected risk at t is given by

$$\{L(t,\boldsymbol{\phi})-R(X_t,t)\}\,b'(t,\boldsymbol{\phi})\Delta t.$$

Hence, the expected risk $f(\boldsymbol{\phi})$ is obtained as follows.

$$\begin{aligned} f(\boldsymbol{\phi}) &= \int_{t0}^{T} E[(L(t,\boldsymbol{\phi})-R(X_t,t))\,b'(t,\boldsymbol{\phi})]\,dt + L(T,\boldsymbol{\phi})(1-P_T(\boldsymbol{\phi})) \\ &= \int_{t0}^{T} E[(R'(X_t,t)-L'(t,\boldsymbol{\phi}))\,b(\int_{t0}^{t}\alpha(X_s,s)\,\phi(X_s,s)\,ds)]\,dt \\ &\quad + L(T,\boldsymbol{\phi}) - E[R(X_T,T)\,b(\int_{t0}^{T}\alpha(X_t,t)\,\phi(X_t,t)\,dt)]. \end{aligned}$$

The condition for the optimal search plan minimizing $f(\boldsymbol{\phi})$ is derived easily by an approach similar to the one mentioned before.

A necessary condition for the R_T-optimal allocation of searching effort is that there exists a non-negative function $\lambda(t)$ such that

$$-A_{yt}^{T}(\boldsymbol{\phi}^*) =(\leq)\ \lambda(t)c_0(y,t) \quad \text{for } \phi^*(y,t) >(=) 0,$$

where

$$A_{yt}^{T}(\boldsymbol{\phi}) = \int_t^T E_{yt}[(R'(X_s,s)-L'(s,\boldsymbol{\phi}))b'(\int_{t0}^{s}\alpha(X_z,z)\phi(X_z,z)dz)] \times q_t(y)\alpha(y,t)ds + c_0(y,t)(1-P_t(\boldsymbol{\phi})) - q_t(y)\alpha(y,t)E_{yt}[R(X_T,T)b'(\int_{t0}^{T}\alpha(X_s,s)\phi(X_s,s)ds)].$$

If $\lambda(t) > 0$, $t \in \mathbf{T}$, then Eq. (2.7) holds.

If $f(\boldsymbol{\phi})$ is convex and $\boldsymbol{\phi}^* \in \boldsymbol{\Psi}_0$, this condition is also a sufficient condition for $\boldsymbol{\phi}^*$ to be optimal.

2.3.2. Optimal Stopping Time of the Search

In this section, we will investigate the optimal stopping time T^* of the search defined by Eq. (2.9). For this purpose, we consider the problem in which the search time is not restricted, and we deal with T as a variable defined on $[t_0, \infty)$ instead of the definition as the limit time of the search in the previous section. T is called the stopping time hereafter.

<u>Lemma 2.5.</u> 1. *If $T \geq T^*$, the R_T-optimal search plan $\boldsymbol{\phi}_T^*$ (conditional on T) that is obtained from Theorem 2.3 is identical to the (unconditionally) optimal search plan $\boldsymbol{\phi}_{T*}^*$.*

(2.14)
$$\phi_T^*(y,t) = \phi_{T*}^*(y,t) \quad \text{if } 0 \leq t \leq T^*,$$
$$\phi_T^*(y,t) = 0 \quad \text{if } T^* < t \leq T.$$

2. *The expected risk $f(\boldsymbol{\phi}_T^*)$ is a nonincreasing function of T and is a constant $f(\boldsymbol{\phi}_{T*}^*)$ for $T > T^*$.*

<u>Proof.</u> 1. From the definition of $\boldsymbol{\phi}_{T*}^*$, $\boldsymbol{\phi}_{T*}^*$ is not influenced by the limit time T of the search if $T \geq T^*$. Therefore, the optimal search plan $\boldsymbol{\phi}_{T*}^*$ is identical to the conditionally optimal search plan $\boldsymbol{\phi}_T^*$ for T ($\geq T^*$).

2. We consider two arbitrary stopping times, T_1 and T_2, $t_0 \leq T_1 < T_2$, and a search plan $\boldsymbol{\phi}_{T2} \in \boldsymbol{\Psi}$ as follows.

$$\phi_{T2}(y,t) = \phi_{T1}^*(y,t) \quad \text{if } t_0 \leq t \leq T_1,$$
$$\phi_{T2}(y,t) = 0 \quad \text{if } T_1 < t \leq T_2.$$

Since $\boldsymbol{\phi}_{T2}$ does not necessary have to be optimal for T_2, $f(\boldsymbol{\phi}_{T1}^*) = f(\boldsymbol{\phi}_{T2}) \geq f(\boldsymbol{\phi}_{T2}^*)$ holds. Hence, $f(\boldsymbol{\phi}_T^*)$ is a nonincreasing function of T in $[t_0, T^*]$. Since $\phi_T^*(y,t) = 0$ for $t \in (T^*, T]$ from Eq. (2.14), $f(\boldsymbol{\phi}_T^*) = f(\boldsymbol{\phi}_{T*}^*)$ for $T \geq T^*$. □

Theorem 2.6. *A necessary condition for T^* to be optimal is*

$$\lambda_T(T) > 0 \quad for \quad T^* - \Delta T < T < T^*,$$
$$\lambda_T(T) = 0 \quad for \quad T^* \leqq T,$$

for a sufficiently small $\Delta T > 0$.

Proof. From the preceding lemma 2.5, the optimal stopping time T^* is given an alternative definition irrespective of Eq.(2.9). The optimal stopping time is the time such that

$$(2.15) \qquad \frac{\partial f(\boldsymbol{\phi}_T^*)}{\partial T} < 0 \qquad \text{for} \quad T^* - \Delta T < T < T^*,$$
$$\frac{\partial f(\boldsymbol{\phi}_T^*)}{\partial T} = 0 \qquad \text{for} \quad T^* \leqq T,$$

for a sufficiently small ΔT.

By considering the variations of $\phi_T^*(y, t)$ and $P_t(\boldsymbol{\phi}_T^*)$ when the stopping time T is prolonged to $T + \Delta T$, the following expression is obtained:

$$\Delta P_t(\boldsymbol{\phi}_T^*) = P_t(\boldsymbol{\phi}_{T+\Delta T}^*) - P_t(\boldsymbol{\phi}_T^*)$$
$$= \int_{t0}^{t} \int_{Z_s} D_{ys}{}^t(\boldsymbol{\phi}_T^*) \Delta\phi_T^*(y, s)\, d\xi(y)\, ds,$$
$$\Delta P_{T+\Delta T}(\boldsymbol{\phi}_{T+\Delta T}^*) = P_{T+\Delta T}(\boldsymbol{\phi}_{T+\Delta T}^*) - P_T(\boldsymbol{\phi}_T^*)$$
$$= \int_{t0}^{T} \int_{Z_s} D_{ys}{}^T(\boldsymbol{\phi}_T^*) \Delta\phi_T^*(y, s)\, d\xi(y)\, ds$$
$$+ \Delta T \int_{Z_T} D_{yT}{}^T(\boldsymbol{\phi}_T^*) \phi_T^*(y, T)\, d\xi(y),$$

where

$$\Delta\phi_T^*(y, t) = \phi_{T+\Delta T}^*(y, t) - \phi_T^*(y, t).$$

Substituting these relations into Eq.(2.8), we have

$$(2.16) \qquad f(\boldsymbol{\phi}_{T+\Delta T}^*) = f(\boldsymbol{\phi}_T^*) + \Delta T\{L'(T, \boldsymbol{\phi}_T^*)(1 - P_T(\boldsymbol{\phi}_T^*))$$
$$- R(T) \int_{Z_T} D_{yT}{}^T(\boldsymbol{\phi}_T^*) \phi_T^*(y, T)\, d\xi(y)\}$$
$$- \int_{t0}^{T} \int_{Z_s} A_{ys}{}^T(\boldsymbol{\phi}_T^*) \Delta\phi_T^*(y, s)\, d\xi(y)\, ds.$$

It is easily derived from Theorem 2.3 that the second term of the right-hand side of Eq.(2.16) is $-\lambda_T(T) C(T) \Delta T$ and the third term is zero. Hence we have

$$(2.17) \qquad \Delta f(\boldsymbol{\phi}_T^*) = f(\boldsymbol{\phi}_{T+\Delta T}^*) - f(\boldsymbol{\phi}_T^*)$$
$$= -\lambda_T(T) C(T) \Delta T.$$

The theorem is directly obtained from the definition of T^* given in Eqs. (2.15) and (2.17). □

A set $\mathbf{S}(T)$ is defined by $\mathbf{S}(T) = \{y \mid \phi_T^*(y, T) > 0\}$. $\mathbf{S}(T)$ is called a search region at T. The following corollary states a property of the search region $\mathbf{S}(T^*)$ at T^*.

Corollary 2.7. *A necessary condition for* $\mathbf{S}(T^*)$ *to be optimal is*

(2.18) $$(1-P_{T*}(\boldsymbol{\phi}_{T*}^*))\int_{S(T*)} c_0(y,T^*)\,d\xi(y) = R(T^*)E_{S(T*)}[b'(\int_{t0}^{T*}\alpha(X_s,s)\phi_{T*}^*(X_s,s)\,ds)].$$

If the stopping time T *is less than* T^*, *we have*

(2.19) $$\lambda_T(T)\int_{S(T)} c_0(y,t)\,d\xi(y)+(1-P_T(\boldsymbol{\phi}_T^*))\int_{S(T)} c_0(y,T)\,d\xi(y) = R(T)E_{S(T)}[b'(\int_{t0}^{T}\alpha(X_s,s)\phi_T^*(X_s,s)\,ds)].$$

Proof. Setting $t = T$ in Eqs.(2.12a),(2.12b) and (2.13), we obtain the following relation on $\mathbf{S}(T)$.

$$\lambda_T(T)c_0(y,T) = R(T)D_{yT}{}^T(\boldsymbol{\phi}_T^*) - c_0(y,T)(1-P_T(\boldsymbol{\phi}_T^*)).$$

Integrating $\lambda_T(T)$ over $\mathbf{S}(T)$, we have

$$\lambda_T(T)\int_{S(T)} c_0(y,T)\,d\xi(y) = R(T)\int_{S(T)} D_{yT}{}^T(\boldsymbol{\phi}_T^*)\,d\xi(y)-(1-P_T(\boldsymbol{\phi}_T^*))\int_{S(T)} c_0(y,T)\,d\xi(y).$$

The first term of the right-hand side of this equation is rewritten as

$$R(T)E_{S(T)}[b'(\int_{t0}^{T}\alpha(X_s,s)\phi_T(X_s,s)\,ds)]$$

from the definition of $D_{yt}{}^T(\boldsymbol{\phi}_T^*)$ given by Eq.(2.13), and therefore, Eq.(2.19) is proved. Substituting T^* for T in Eq.(2.19), we have Eq.(2.18) since $\lambda_{T*}(T^*) = 0$ by Theorem 2.6. □

Remark 2.6. By the definition, the optimal stopping time T^* is such a time that the search is conducted in $[T^*-\Delta T, T^*]$ and is stopped eventually in (T^*,∞). As described before, $\lambda(t)c_0(y,t)$ is the expected marginal return under the R_T-optimal plan, and since $c_0(y,T) > 0$, the inequality $\lambda(T) > 0$ means that the expected marginal return is positive. Therefore, the meaning of Theorem 2.6 is as follows. If the expected marginal return at T is positive, the search should be continued because the search is motivated by the positive increment of the expected return, and the reverse is also true. On the contrary, if the search should be stopped at T, $\lambda(T)$ must not be positive. Since $\lambda(t)$, $t \in \mathbf{T}$, is non-negative in the R_T-optimal plan, $\lambda(T) = 0$ for $T \geq T^*$.

Remark 2.7. It is important to distinguish the limit time T of the search in the previous section and the stopping time in Theorem 2.6. As mentioned in the preceding remark 2.6, the stopping time T^* implies that the search is conducted at $T^*-\Delta T$ and is stopped at T^*. On the other hand, the limit time T means merely an upper limit of the available search time and if $T > T^*$ the search is not to be conducted in the interval $(T^*, T]$ as stated in Lemma 2.5. Therefore, Theorem 2.3 includes Lemma 2.5 in this case.

2.3.3. A Numerical Example

In the previous sections, we derived the theorems under the assumption of continuous time. However the continuous time space is changed easily with the discrete time point system. A model in which both the time space and the target space are discrete is of special importance when we calculate the optimal search plan numerically. In this model, the theorems derived earlier are valid if the integrations with respect to t and y are interchanged with the summations.

In this section, we will examine the following numerical example. Suppose the target space consists of discrete regions named boxes, numbered $j = 1, 2, \cdots, 5$ from the left to the right, and the time space is discrete time points t, $t = 1, 2, \cdots$. We assume that if the target is in box j ($j \leq 4$) at t, the target selects either j or $j+1$ with probability (0.5, 0.5) for the next position and if box 5 is chosen, the target stays there forever. Namely, the transition probaoility matrix $[p(j, k)]$ is given by

$$[p(j,k)] = \begin{bmatrix} 0.5 & 0.5 & 0 & 0 & 0 \\ 0 & 0.5 & 0.5 & 0 & 0 \\ 0 & 0 & 0.5 & 0.5 & 0 \\ 0 & 0 & 0 & 0.5 & 0.5 \\ 0 & 0 & 0 & 0 & 1 \end{bmatrix}.$$

The initial distribution of the target is assumed to be uniform over j; $q_1(j) = 0.2$, $j = 1, 2, \cdots, 5$. It is also assumed that the search time is limited to $T = 10$ and a total search cost $C(t) = 7.5$ is available to the searcher at each time. The conditional detection function $b(\varphi)$ is assumed to be an exponential function $b(\varphi) = 1 - \exp(-\varphi)$. The unit cost of the search effort is $c_0(j, t) = 5$ and the reward given when the target is detected successfully is $R(t) = 100$. For simplicity, these values, $p(j, k)$, $C(t)$, $c_0(j, t)$, and $R(t)$, are assumed to be constant during the whole search time. As mentioned in Remark 2.4, the FAB algorithm proposed by Washburn (1983) is applied and we obtain the R_T-optimal allocation of searching effort as shown in Table 2.1. In Table 2.1, we observe that the R_T-optimal search plan is started with $\lambda(1)=0$ (called a partial search), followed by a break at $t = 2$, 3, and 4. After that, the partial search resumes, and finally $\lambda(t) > 0$ at $t = 8$, 9, and 10 (the complete search) in which the effort is concentrated to the right-most box takes place. This result is explained intuitively as follows. Because of the Markov chain-like motion of the target, the target is more and more likely in the right-most box as time passes. Therefore, if the limit time of the search is long enough, the searcher

Table 2.1. $\phi_{10}^*(j, t)$.

$t\backslash j$	1	2	3	4	5	Total	$\lambda(t)$
1	0.415	0	0	0	0	0.415	0
2	0	0	0	0	0	0	0
3	0	0	0	0	0	0	0
4	0	0	0	0	0	0	0
5	0	0	0	0	0.165	0.165	0
6	0	0	0	0	0.420	0.420	0
7	0	0	0	0	0.776	0.776	0
8	0	0	0.212	0.290	0.998	1.500	0.071
9	0	0	0.027	0.454	1.018	1.500	0.716
10	0	0	0	0.265	1.235	1.500	1.270
	$f(\boldsymbol{\phi}_{10}^*)$			-82.822			

would wait all the time and put his full effort into box 5 at the last time point. If the limit time of the search is shortened a little, its effect might be this: First, even at the last time point, there is some probability that the target is still in box 4 or 3, and thus the searcher has to divide his effort to these boxes besides box 5, the most probable. The effort distribution at t = 8, 9, and 10 in Table 2.1 is the case. The second effect is on starting the partial search. Calculation shows that if the target was in box 1 at t = 1, the probability of it still being in boxes 1 through 4 at t = 10 is about 0.254 (=$1-p^{(10)}(1,5)$, where $p^n(i, j)$ is the n-step transition probability given by $[p^n(i, j)] = [p(i, j)]^n$.) To reduce the probability, searching box 1 at the beginning might be effective, which explains the partial search at t = 1. If the limit time of the search is shortened further, the two effects grow and the break of search shrinks and disappears, as is seen in Tables 2.2 and 2.3. It is interesting to note that the allocation of searching effort at the end of the search is identical irrespective of T:

$$\phi_T^*(4, T) = 0.265, \quad \phi_T^*(5, T) = 1.235 \quad \text{for } T = 3,\ 6, \text{ and } 10.$$

Table 2.2. $\phi_6^*(j, t)$.

$t\backslash j$	1	2	3	4	5	Total	$\lambda(t)$
1	0.995	0.364	0.144	0	0	1.463	0
2	0	0.327	0.332	0.090	0.229	0.978	0
3	0	0.233	0.258	0.197	0.667	1.355	0
4	0	0.144	0.353	0.291	0.712	1.500	0.522
5	0	0	0.329	0.362	0.809	1.500	1.085
6	0	0	0	0.265	1.235	1.500	1.712
	$f(\boldsymbol{\phi}_{10}^*)$			-69.689			

Table 2.3. $\phi_3^*(j, t)$.

$t\backslash j$	1	2	3	4	5	Total	$\lambda(t)$
1	0.769	0.475	0.256	0	0	1.500	4.089
2	0	0.378	0.514	0.292	0.316	1.500	4.142
3	0	0	0	0.265	1.235	1.500	5.508
$f(\boldsymbol{\phi}_{10}^*)$				-49.725			

The expected risk $f(\boldsymbol{\phi}_T^*)$ of the R_T-optimal search plan is shown in Fig. 2.1. As mentioned before, $f(\boldsymbol{\phi}_T^*)$ is a strictly decreasing function of T since the target density concentrates to box 5 as time passes. Therefore, no stopping is optimal in this case. This is also suggested by the fact that the $\lambda_T(t)$s in Tables 2.1~2.3 increase with t.

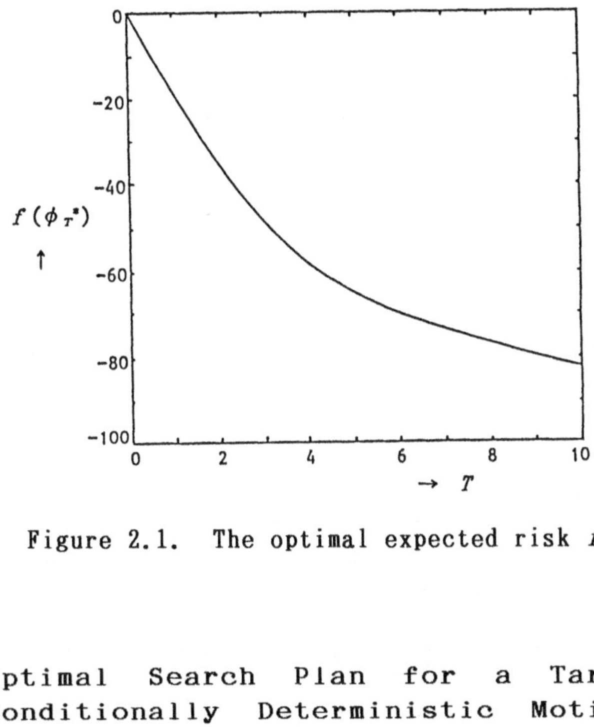

Figure 2.1. The optimal expected risk $f(\boldsymbol{\phi}_T^*)$.

2.4. Optimal Search Plan for a Target with Conditionally Deterministic Motion

In this section, we will investigate an optimal search problem for a moving target with a rather simple type of motion called factorable conditionally deterministic. "Conditionally deterministic" means that the target motion depends on stochastic parameters and if the parameters are known, the target position is determined at all time in the future. "Factorable" means the target motion has the properties that the Jacobian of the transformation that relates the target motion parameters to the target position is separable into two factors, time-dependent and space-dependent parts. The target with these two properties is called the target with factorable conditionally deterministic motion. We call it the F.C.D.M. target.

As mentioned earlier, problems of optimal search for a moving target have been studied by many authors and generalized theorems have been obtained. Therefore, from the theoretical viewpoint, the search problem for the F.C.D.M. target is now less interesting, but the class of search for the F.C.D.M. target contains many important cases for real-world applications; i.e., the problem is still worth investigating. Moreover, up to now, the studies of the optimal search for the F.C.D.M. target have been limited to problems maximizing the detection probability subject to a constraint on the searching effort rate. We call it the P-optimal search problem. In this section, we investigate the optimal search plan that minimizes the expected risk subject to a constraint on the searching effort rate. We call this the R-optimal search problem. We derive the R-optimal search plan for the F.C.D.M. target moving diffusively from a fixed point explicitly. This section is taken from Iida (1989b). The detailed description of the search problem dealt with here follows:

1. The target movement and search operation take place in an Euclidean n-space $\mathbf{Y}$ and in a continuous time interval $\mathbf{T} = [t_0, T]$, $0 \leq t_0 < T < \infty$.

2. The target motion is assumed to be characterized by a stochastic parameter X that has a probability density function $p(x)$ defined on a Euclidean n-space $\mathbf{X}$. To prescribe the target motion, we define a Borel function $\eta : \mathbf{X} \times \mathbf{T} \rightarrow \mathbf{Y}$. Here, if $X = x$, the position of the target $y \in \mathbf{Y}$ at $t \in \mathbf{T}$ is given by $y = \eta(x, t)$. For $t \in \mathbf{T}$, let η_t be the transformation $\eta(\cdot, t)$ and η_t^i be the ith component of η_t. $\eta_t(x)$ is assumed to have the continuous partial derivative at all $x \in \mathbf{X}$ and we denote the partial derivative of η_t^i with respect to the kth component of x by $\partial \eta_t^i(x)/\partial x_k$. Let $J(x, t)$ be the absolute value of the Jacobian of the transformation $\eta_t(x)$ evaluated at x.

$$J(x, t) = \left| \det\left(\frac{\partial \eta_t^i(x)}{\partial x_k} \right) \right| \quad \text{for } x \in \mathbf{X},\ t \in \mathbf{T}.$$

The following properties are assumed. (1) η_t is a one-to-one correspondence for $t \in \mathbf{T}$; and (2) $J(x, t)$ is positive for all $x \in \mathbf{X}$ and $t \in \mathbf{T}$. By these assumptions, the C.D.M. target is defined precisely. The inverse map of η_t, $\eta_t^{-1}: \mathbf{Y} \rightarrow \mathbf{X}$ can be defined on the image of η_t from the first assumption and the integration defined on $\mathbf{Y}$ is transformed to that on $\mathbf{X}$ by the second. Hence, the probability of the target being in a Borel set $\mathbf{S} \subseteq \mathbf{Y}$ at t is given by $\int_{\{\eta_t^{-1}(s)\}} p(x)dx$. Furthermore, the target motion is called factorable, if there exist Borel functions $j: \mathbf{X} \to (0, \infty)$ and $n: \mathbf{T} \to (0, \infty)$ such that (3) $J(x, t) = j(x)n(t)$ for $\{x \in \mathbf{X}, t \in \mathbf{T}\}$, in addition to the two previous assumptions. Here we assume $n'(t) > 0$. This means that the target moves diffusively.

3. The search is started at t_0, $0 \leq t_0 < \infty$, and the searcher can stop the search at any time he wants. Total searching effort rate $\Phi(t)$ is available to the searcher at each t, and $\{\Phi(t), t \geq t_0\}$ is given before the start of the search. We assume that $\Phi(t)$ is continuously divisible in the target space.

4. Let $\phi(y, t)$ be a density of searching effort rate allocated to y at t. A set of $\phi(y, t)$ is denoted by $\boldsymbol{\phi} = \{\phi(y, t), y \in \mathbf{Y}, t \in \mathbf{T}\}$.

5. If a motion parameter x is selected by the target and an allocation of searching effort density $\boldsymbol{\phi}$ is employed by the searcher, the cumulative searching effort density to this target until t is given by $\psi = \int_{t_0}^{t} \phi(\eta_s(x), s)\,ds$. A detection function $b(x, \psi)$ gives the conditional probability of detecting the target until t. $b(x, \psi)$ is assumed to be regular with respect to ψ and the derivative of $b(x, \psi)$ with respect to ψ is denoted by $b'(x, \psi)$.

6. Some reward $R(t)$ is given to the searcher when he detects the target at t. $R(t)$ is assumed to be a positive, nonincreasing, and differentiable function of t; $0 < R(t) < \infty$ and $R'(t) \leq 0$.

7. The unit cost of the searching effort density at t is given by $c_0(t)$ (>0), $t \in \mathbf{T}$. $c_0(t)$ is assumed to be a positive, nondecreasing, and differentiable function of t; $0 < c_0(t) < \infty$ and $c_0'(t) \geq 0$.

8. In our problem, we assume $R'(t) \leq 0$, $c_0'(t) \geq 0$, and $n'(t) > 0$ (the region covers the possible position of the target expands as time passes). These three assumptions are more restrictive than those given in Section 2.3. Then, since the search becomes more expensive with time, the searcher should stop the search when an unsuccessful search is continued too long. Therefore, we must consider two factors of the search plan; the allocation of searching effort $\boldsymbol{\phi} = \{\phi(y, t), y \in \mathbf{Y}, 0 \leq t \leq T\}$ and the stopping time T of the search. A decision pair $(\boldsymbol{\phi}, T)$ is called a search plan. The measure of effectiveness of a search plan is assumed to be the expected risk of the search.

The formulation of the problem is as follows. We consider a conditional search plan $\boldsymbol{\phi}_T$ with a stopping time T. (Sometimes the subscript T in $\boldsymbol{\phi}_T$ is omitted, if confusion is not expected.) Since the detection function is given by $b(x, \psi)$, the detection probability of the target until t, $P_t(\boldsymbol{\phi})$, is obtained by

$$(2.20) \qquad P_t(\boldsymbol{\phi}) = \int_X p(x)\, b\Big(x, \int_{t_0}^{t} \phi(\eta_s(x), s)\,ds\Big)\,dx.$$

The cumulative searching cost $L(t, \boldsymbol{\phi})$ up to t is given by

$$(2.21) \qquad L(t, \boldsymbol{\phi}) = \int_{t_0}^{t} c_0(s)\, n(s) \Big\{\int_X \phi(\eta_s(x), s)\, j(x)\,dx\Big\}\,ds.$$

Differentiation of Eq.(2.21) with respect to t yields

$$(2.22) \qquad L'(t, \boldsymbol{\phi}) = \frac{\partial L(t, \boldsymbol{\phi})}{\partial t} = c_0(t)n(t)\int_X \phi(\eta_t(x), t)j(x)dx.$$

The expected risk $f(\boldsymbol{\phi})$ is obtained by Eqs.(2.20)～(2.22) as

$$(2.23) \qquad f(\boldsymbol{\phi}) = \int_{t_0}^{T}\{R'(t)-L'(t, \boldsymbol{\phi})\}P_t(\boldsymbol{\phi})dt + L(T, \boldsymbol{\phi}) - R(T)P_T(\boldsymbol{\phi}).$$

On the other hand, the total searching effort density is constrained by $\Phi(t)$, the searching effort density $\phi(y, t)$ is nonnegative, and the stopping time T is larger than t_0. By the assumptions, these restrictions are written as follows:

$$(2.24) \qquad \begin{aligned} &n(t)\int_X \phi(\eta_t(x), t)j(x)dx \leqq \Phi(t) \quad \text{for all } t \in \mathbf{T}, \\ &\phi(\eta_t(x), t) \geqq 0 \quad \text{for all } x \in \mathbf{X} \text{ and } t \in \mathbf{T}, \\ &T \geqq t_0. \end{aligned}$$

Let $\boldsymbol{\Psi}$ be the set of search plans that satisfies the constraints; we pay attention only to the search plan of the class $\boldsymbol{\Psi}$. $\boldsymbol{\Psi}_0$ is defined as the set of $\boldsymbol{\phi} \in \boldsymbol{\Psi}$ such that

$$n(t)\int_X \phi(\eta_t(x), t)j(x)dx = \Phi(t).$$

Therefore, the problem is formulated as a variational problem to find a search plan $(\boldsymbol{\phi}, T)$ that minimizes the functional $f(\boldsymbol{\phi})$ given by Eq.(2.23) subject to the restrictions in Eq.(2.24). It should be noted that the problem of searching for the F.C.D.M. target in space $\mathbf{Y}$ is converted to that of a search for a stationary target in space $\mathbf{X}$.

The conditionally R-optimal allocation of searching effort is obtained by relating it to the P-optimal allocation. First, we examine the properties of the P-optimal allocation.

If there exists $\boldsymbol{\phi}^0 \in \boldsymbol{\Psi}$ such that

$$P_t(\boldsymbol{\phi}^0) = \max\{P_t(\boldsymbol{\phi}), \boldsymbol{\phi} \in \boldsymbol{\Psi}\} \quad \text{for all } t \in \mathbf{T},$$

then $\boldsymbol{\phi}^0$ is called uniformly P-optimal within $\boldsymbol{\Psi}$. When the target is stationary, the uniformly P-optimal allocation always exists and the properties of the plan have been studied by many authors. However, the uniformly P-optimal plan for a moving target does not always exist. Here, we consider a maximization problem of the objective function $P_t(\boldsymbol{\phi})$ given by Eq.(2.20) subject to the constraints in Eq.(2.24). This problem is formally nothing but a search problem for a stationary target in the motion parameter space $\mathbf{X}$, and therefore, there exists a uniformly P-optimal plan if the target motion is F.C.D.M. The uniformly P-optimal allocation was obtained by Stone and Richardson (1974). The properties of the uniformly P-optimal allocation of searching effort for the F.C.D.M. target are summarized from Stone (1975) as follows.

<u>Lemma 2.8.</u> *Suppose the target motion is F.C.D.M. and* $b(x,\varphi)$ *is regular on* $\mathbf{X}$. *Let* $\psi(x,t)$ *be*

(2.25) $$\psi(x,t) = \int_{t_0}^{t} \phi(\eta_s(x), s)\,ds.$$

1. *A necessary and sufficient condition for* $\boldsymbol{\phi}^0$ *to be uniformly P-optimal is that there exists a nonnegative function* $\mu(t)$ *such that for a.e.* $t \in \mathbf{T}$ *and* $x \in \mathbf{X}$,

(2.26a) $$\frac{p(x)\,b'(x,\psi^0(x,t))}{j(x)} = \mu(t) \qquad \text{if } \phi^0(x,t) > 0,$$

(2.26b) $$\frac{p(x)\,b'(x,0)}{j(x)} \leqq \mu(t) \qquad \text{if } \phi^0(x,t) = 0,$$

where $\mu(t)$ *is determined from the following constraints:*

(2.27) $$n(t)\int_X \phi^0(x,t)\,j(x)\,dx = \Phi(t).$$

2. *The uniformly P-optimal allocation is a member of* $\boldsymbol{\Psi}_0$, $\boldsymbol{\phi}^0 \in \boldsymbol{\Psi}_0$.
3. $\mu(t)$ *is a strictly decreasing function of t.*

We can apply Theorem 2.3 to our problem and obtain the following theorem of the R-optimal allocation of searching effort for the diffusive F.D.C.M. target. Here, as mentioned before, the system parameters are assumed as

(2.28) $$0 < c_0(t) < \infty, \quad c_0'(t) \geqq 0, \quad 0 < R(t) < \infty, \quad R'(t) \leqq 0, \quad n'(t) > 0.$$

<u>Theorem 2.9.</u> *If an inequality*

(2.29) $$Q_T(\boldsymbol{\phi}^0) < \frac{R(T)\,\mu(T)}{c_0(T)\,n(T)}$$

holds, the uniformly P-optimal allocation of searching effort $\boldsymbol{\phi}^0$ *given by Eqs.*(2.26a) *and* (2.26b) *is also* R_T*-optimal, namely* $\boldsymbol{\phi}_T^* = \boldsymbol{\phi}^0$.

<u>Proof.</u> The theorem can be proved by showing that the uniformly P-optimal allocation of searching effort $\boldsymbol{\phi}^0$ given by Lemma 2.8 satisfies the conditions for the R_T-optimal search plan given by Theorem 2.3.

Consider the uniformly P-optimal search plan $\boldsymbol{\phi}^0$, and we have

$$D_{yt}^{s}(\boldsymbol{\phi}^0) = \frac{p(x)\,b'(x,\psi^0(x,s))}{n(t)\,j(x)} \qquad \text{for } x = \eta_t^{-1}(y),$$
$$L'(s,\boldsymbol{\phi}^0) = c_0(s)\,\Phi(s),$$
$$P_t(\boldsymbol{\phi}^0) = \int_X p(x)\,b(x,\psi^0(x,t))\,dx,$$

by Lemma 2.8. From these relations and Eq.(2.13), we obtain

(2.30) $$-A_{yt}^{T}(\boldsymbol{\phi}^0) = \int_t^T (c_0(s)\,\Phi(s) - R'(s))\,p(x)\,b'(x,\psi^0(x,s))\,\frac{1}{j(x)\,n(t)}\,ds - c_0(t)\,Q_t(\boldsymbol{\phi}^0) + R(T)\,p(x)\,b'(x,\psi^0(x,T))\,\frac{1}{j(x)\,n(t)},$$

where $x = \eta_t^{-1}(y)$.

Here, we consider the case $\phi^0(y,t) > 0$. In this case, since $\phi^0(x,t) = \phi^0(\eta_t^{-1}(y), s) > 0$ for all $s \in [t,T]$, $p(x)b'(x, \phi^0(x,s))/j(x) = \mu(s)$ holds for all $s \in [t,T]$ by Eqs.(2.26a) and (2.26b), and therefore, Eq.(2.30) is rewritten as

$$-A_{y,t}{}^T(\boldsymbol{\phi}^0) = \int_t^T (c_0(s)\Phi(s) - R'(s)) \frac{\mu(s)}{n(t)} ds - c_0(t) Q_t(\boldsymbol{\phi}^0) + R(T)\frac{\mu(T)}{n(t)}.$$

Since the right-hand side of this equation does not depend on x and is a function of t and T, we denote it as $\lambda_T(t)$.

$$\lambda_T(t) = \int_t^T (c_0(s)\Phi(s) - R'(s)) \frac{\mu(s)}{n(t)} ds - c_0(t) Q_t(\boldsymbol{\phi}^0) + R(T)\frac{\mu(T)}{n(t)}. \tag{2.31}$$

Then, by the condition of Eq.(2.29), we have

$$\lambda_T(T) = -c_0(T) Q_T(\boldsymbol{\phi}^0) + R(T)\frac{\mu(T)}{n(T)} > 0. \tag{2.32}$$

Since $\partial P_T(\boldsymbol{\phi}^0)/\partial t = \Phi(t)\mu(t)/n(t)$ from Eqs.(2.20),(2.24),(2.26a) and (2.26b), we obtain

$$\frac{d\lambda_T(t)}{dt} = R'(t)\frac{\mu(t)}{n(t)} + \frac{n'(t)}{n(t)^2}\int_t^T (R'(s) - c_0(s)\Phi(s))\mu(s) ds - c_0{}'(t) Q_t(\boldsymbol{\phi}^0) - \frac{R(T)\mu(T)n'(t)}{n(t)^2} < 0$$

from the assumptions in Eq.(2.28). Therefore, $\lambda_T(t)$ is a strictly decreasing function of t and $\lambda_T(t) > 0$ for all $t \in [t_0, T]$ by Eq.(2.32). Hence, $\boldsymbol{\phi}^0 \in \boldsymbol{\Psi}_0$ for all $t \in [t_0, T]$ from Theorem 2.3.

Next we will consider the case $\phi^0(y,t) = 0$. Since $\phi^0(\eta_t^{-1}(y), s) \geq 0$ in the interval $s \in [t,T]$,

$$\left.\frac{p(x)b'(x, \phi^0(x,t))}{j(x)}\right|_{x=\eta_t^{-1}(y)} \leq \mu(t)$$

in Eq.(2.30) by Eqs.(2.26a) and (2.26b). Then $-A_{y,t}{}^T(\boldsymbol{\phi}^0) \leq \lambda_T(t)$ is derived from Eqs.(2.30) and (2.31). Therefore, the uniformly P-optimal allocation ϕ^0 given by Lemma 2.8 is the R_T-optimal allocation given by Theorem 2.3. □

The conditionally optimal allocation of searching effort $\boldsymbol{\phi}_T{}^*$ is called additive if it has the following property. Suppose the conditionally optimal allocation $\boldsymbol{\phi}_{T1}{}^*$ with a stopping time T_1 is employed and it fails to detect the target. At T_1, the searcher is allowed to add the conditionally optimal allocation $\boldsymbol{\phi}_{T2}{}^*$ with a stopping time T_2 (from T_1) given that the search $\boldsymbol{\phi}_{T1}{}^*$ fails. On the other hand, we consider the case in which the stopping time $T = T_1 + T_2$ is given in advance and the optimal search plan $\boldsymbol{\phi}_T{}^*$ with stopping time T is employed. Then if ϕ_T^* is identical with the allocation $\{\phi_{T1}^*, \phi_{T2}^*\}$;

$$\phi_T^*(y,t) = \phi_{T1}^*(y,t) \quad \text{for} \quad t_0 \leqq t \leqq T_1,$$
$$\phi_T^*(y,t) = \phi_{T2}^*(y,t) \quad \text{for} \quad T_1 < t \leqq T_1+T_2,$$

the allocation ϕ_T^* is called additive and is denoted as $\phi_T^* = \phi_{T1}^* + \phi_{T2}^*$. The following corollary is derived easily from Theorem 2.9.

Corollary 2.10. *The conditionally R_T-optimal allocation of searching effort determined by Theorem* 2.9 *is uniformly R-optimal and has the additivity.*

Proof. Theorem 2.9 states that the R_T-optimal allocation of searching effort ϕ_T^* given T is identical with the uniformly P-optimal allocation ϕ^0. Let S be an arbitrary stopping time, $t_0 \leqq S \leqq T$. Then the R_S-optimal allocation ϕ_S^* is also equal to ϕ^0. Since the uniformly P-optimal allocation ϕ^0 is not affected by the stopping time, ϕ_S^* is also identical with ϕ_T^* in $[t_0, S]$. Since S is arbitrary in $[t_0, T]$, this implies that the R_S-optimal allocation ϕ_S^* minimizes the risk of the search at any time in $[t_0, T]$ (the uniformity of ϕ_T^*) and ϕ_T^* is identical with $\phi_S^* + \phi_{T-S}^*$ (the additivity of ϕ_T^*). □

The following theorem is useful to determine the optimal stopping time T^* of the search for the diffusive F.C.D.M. target.

Theorem 2.11. *We assume the inequalities in Eq.*(2.28).

1. *If an inequality*

(2.33)
$$Q_T(\phi_T^*) < \frac{R(T)\mu(T)}{c_0(T)n(T)}$$

is satisfied, the search should not be stopped at T, where

$$Q_T(\phi) = 1 - P_T(\phi).$$

2. *If $R(t_0)\mu(t_0)/c_0(t_0)n(t_0) > 1$ and $d\lambda_T(T)/dT \leqq 0$, and if an equation $Q_T(\phi_T^*) = R(T)\mu(T)/c_0(T)n(T)$ has nonnegative solutions, the optimal stopping time T^* is given by*

(2.34)
$$T^* = \min\left\{T \mid Q_T(\phi_T^*) = \frac{R(T)\mu(T)}{c_0(T)n(T)}\right\}.$$

If the equation has no solution, the search should not be stopped until detection.

Proof. 1. Since the inequalities in Eqs.(2.28) and (2.33) are assumed, Theorem 2.9 is established and $\phi_T^* = \phi^0$, therefore, $\lambda_T(T)$ is given by

(2.35)
$$\lambda_T(T) = -c_0(T)Q_T(\phi_T^*) + \frac{R(T)\mu(T)}{n(T)}$$

from Eq.(2.31). The value $\lambda_T(T)$ is positive under Eq.(2.33), and therefore, stopping at T is not optimal by Theorem 2.6.

2. $R(t_0)\mu(t_0)/c_0(t_0)n(t_0) > 1$ is equivalent to $\lambda_{t0}(t_0) > 0$ and $Q_T(\phi_T^*) = R(T)\mu(T)/c_0(T)n(T)$ in Eq.(2.34) means $\lambda_T(T) = 0$ from Eq. (2.35). If $\lambda_T(T) = 0$ has a nonnegative solution T^* given by Eq.(2.34), $\lambda_T(T)$ is positive in the interval $t_0 \leqq T < T^*$ and $\lambda_T(T)$ is zero for all $T \geqq T^*$, since $\lambda_T(T) \geqq 0$ by the definition of $\lambda_T(T)$ and $\lambda_T(T)$ is nonincreasing by the condition of the theorem. Hence, the theorem is proved. If $\lambda_T(T) = 0$ has no solution, the search should be continued until detection since $\lambda_T(T) > 0$ for all T. □

If the condition of the following corollary holds, the search should not be started at all.

Corollary 2.12. *We assume the condition of Theorem* 2.11. *Let* $\bar{\mu}$ *be*

$$\bar{\mu} = \sup_x \left\{ \frac{p(x)\,b'(x,0)}{j(x)},\ x \in X \right\}.$$

If $\bar{\mu} R(t_0)/c_0(t_0)n(t_0) < 1$, *then the search should not be started.*

Proof. Differentiating $f(\phi_T^*)$ with respect to T, we obtain

$$\frac{\partial f(\phi_T^*)}{\partial T} = \Phi(T)\left\{ c_0(T)Q_T(\phi_T^*) - \frac{R(T)\mu(T)}{n(T)} \right\},$$

because $L'(T, \phi_T^*) = c_0(T)\Phi(T)$ and $\partial P_T(\phi_T^*)/\partial T = \Phi(T)\mu(T)/n(T)$ from Lemma 2.8 and Theorem 2.9. Since $\mu(T) \leqq \bar{\mu}$ by Lemma 2.8 for all T, $\partial f(\phi_T^*)/\partial T$ is positive for all T under the condition of the corollary. Therefore, the search should not be started at all. □

Next, we consider a search situation in which a searcher detects a target at $t = 0$, and in a moment, the target begins its evasion. The searcher loses contact with the target soon after the beginning of evasion, and he must search to relocate the target (called the datum search). It is assumed that some time lag t_0 (>0) elapses before the start of the datum search and the total searching effort rate $\Phi(t)$ at each time t is available to the searcher. The target chooses its course θ at random over $[0, 2\pi]$ uniformly and his speed u with probability density function $g(u)$, $u \geqq 0$, and thereafter keeps the course and speed. We assume that both the position of the target at $t = 0$ and the density function $g(u)$ is known to the searcher. Without any loss of generality, the initial position of the target is taken to be the origin of the target space. In the following, we derive the R-optimal search plan for the F.C.D.M. target. New assumptions of the search model are:

1. The target space is a two-dimensional continuous space.
2. The conditional detection function is assumed to be exponential,

$$b(x, \varphi(x,t)) = 1 - \exp(-\varphi(x,t)).$$

3. We consider four cases of $g(u)$, the uniform distribution, the triangular distribution, the Rayleigh distribution and the gamma distribution. Other assumptions are the same as before.

The polar coordinate system (r, θ) is employed and the target motion parameter is defined by $x = (u, \theta)$. The function $\eta_t(x)$ is given by

$$y = \eta_t(x); \ (r\cos\theta,\ r\sin\theta) = (ut\cos\theta,\ ut\sin\theta).$$

Then, $J(x, t) = ut^2$ is obtained. Hence

$$j(x) = u, \qquad n(t) = t^2.$$

Meanwhile, the probability density function of x is obtained from the assumption as

$$p((u, \theta)) = \frac{g(u)}{2\pi}.$$

Therefore, from Lemma 2.8 and Theorem 2.9, the R-optimal allocation $\phi^*(x, t)$ satisfies the condition: if $\phi^*((u, \theta), t) > 0$,

$$\frac{g(u)\exp\{-\varphi^*((u, \theta), t)\}}{2\pi u} = \mu(t).$$

This yields

(2.36) $$\varphi^*((u, \theta), t) = \log \frac{g(u)}{2\pi\mu(t)u} \quad \text{if } g(u) > 2\pi\mu(t)u,$$
$$\varphi^*((u, \theta), t) = 0 \quad \text{otherwise.}$$

Let $\mathbf{U}_t = \{u | g(u) > 2\pi\mu(t)u\}$, and Eq. (2.27) becomes

(2.37) $$\int_{U_t} u \log \frac{g(u)}{2\pi\mu(t)u}\, du = \int_{t_0}^{t} \frac{\Phi(s)}{2\pi s^2}\, ds \equiv \Theta(t).$$

From Eqs. (2.25), (2.36) and $\phi^*((u, \theta), t) = \partial\varphi^*((u, \theta), t)/\partial t$, we obtain

$$\phi^*((u, \theta), t) = \frac{d(-\log\mu(t))}{dt} \quad \text{for } u \in \mathbf{U}_t,$$
$$\phi^*((u, \theta), t) = 0 \quad \text{otherwise.}$$

From this relation, the R-optimal allocation of searching effort on the target space is obtained by

$$\phi^*((r, \theta), t) = \frac{d(-\log\mu(t))}{dt} \quad \text{if } g(\tfrac{r}{t}) \geq \frac{2\pi\mu(t)r}{t},$$
$$\phi^*((r, \theta), t) = 0 \quad \text{otherwise.}$$

It should be noted that in this equation, $\phi^*((r, \theta), t)$ does not depend on (r, θ), and therefore, $\phi^*((r, \theta), t)$ is distributed uniformly in a region $\mathbf{S}(t)$ defined by

(2.38) $$\mathbf{S}(t) = \{(r, \theta) | g(\tfrac{r}{t}) \geq \frac{2\pi\mu(t)r}{t}\}.$$

$\mathbf{S}(t)$ is the region of the target space in which the searching effort is allocated at t. Hence, $\phi^*((r, \theta), t)$ is denoted by

$$(2.39)\qquad \phi^*((r,\theta),t) = \frac{\Phi(t)}{\|\mathbf{S}(t)\|} \quad \text{for } (r,\theta) \in \mathbf{S}(t),$$
$$\phi^*((r,\theta),t) = 0 \quad \text{otherwise},$$

where $\|\mathbf{S}(t)\|$ is the area of the region $\mathbf{S}(t)$.

Next, we will derive the optimal stopping time T^*. Substituting $\overline{\mu} = \max_u(g(u)/2\pi u)$ into the condition of Corollary 2.12, we obtain the following condition, in which the search should not be started:

$$(2.40)\qquad \frac{R(t_0)}{c_0(t_0)t_0^2}\max_u\left(\frac{g(u)}{2\pi u}\right) < 1.$$

On the other hand, if the inequality sign is reversed in Eq. (2.40), the search is started and the optimal stopping time T^* is determined by Eq. (2.34). Since $Q_T(\boldsymbol{\phi}_T{}^*)$ is given by $\int g(u)\exp(-\psi(u,T))\,du$, T^* is obtained from

$$(2.41)\qquad \int_{\gamma_{T^*}{}^{-1}(\mathbf{S}(T^*))} g(u)\exp(-\psi^*(u,T^*))\,du = \frac{R(T^*)\,\mu(T^*)}{c_0(T^*)\,T^{*2}}.$$

If we specify the density function $g(u)$, the R-optimal allocation $\boldsymbol{\phi}_T{}^*$ is obtained from Eq. (2.39) explicitly, and then we can easily prove that $\lambda_T(T)$ is a decreasing function of T. In this case, since Theorem 2.11 is applicable, the optimal stopping time T^* is determined by Eq. (2.41). Here, we shall investigate the R-optimal search plan for four density functions $g(u)$ important to real-world applications. In many cases of the actual search situation, the reward and searching cost are constant, therefore, we assume $R(t) = R$ and $c_0(t) = c_0$ from now on.

1. Uniform Distribution on $[u_0, u_1]$

Let $g(u)$ be the uniform distribution over $[u_0, u_1]$,

$$g(u) = \frac{1}{u_1 - u_0} \quad \text{for } u_0 \le u \le u_1,$$
$$g(u) = 0 \quad \text{otherwise}.$$

Since $\overline{\mu} = 1/(2\pi u_0(u_1-u_0))$, the searcher should not begin the search if

$$\frac{R}{2\pi u_0 c_0 (u_1-u_0) t_0^2} < 1$$

from Eq. (2.40). If the inequality sign is reversed here, the search should be started and the R-optimal search plan $\boldsymbol{\phi}^*$ is obtained as follows. By the definition

$$\mathbf{U}_t = \{u \mid u_0 \le u \le u(t),\ u(t) = \frac{1}{2\pi\,\mu(t)(u_1-u_0)}\},$$

Eq. (2.37) is rewritten as

$$(2.42)\qquad \left(\frac{u(t)}{u_0}\right)^2 - \log\left(\frac{u(t)}{u_0}\right)^2 - 1 = \frac{4\,\Theta(t)}{u_0^2}.$$

The solution $u(t)$ (normalized by u_0) to Eq.(2.42) is shown in Fig. 2.2 as a function of $4\Theta(t)/u_0^2$.

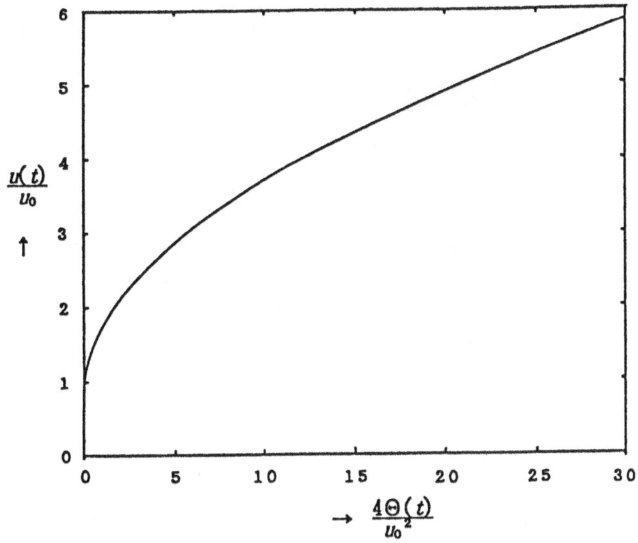

Figure 2.2. Solution $(u(t)/u_0)$ to Eq.(2.42).

If the value $\Theta(t)$ defined by Eq.(2.37) satisfies

$$0 \leqq \Theta(t) \leqq \frac{u_0^2}{4}\{(\frac{u_1}{u_0})^2 - \log(\frac{u_1}{u_0})^2 - 1\},$$

Eq.(2.42) has a unique solution $u(t)$, $u_0 \leqq u(t) \leqq u_1$. Then the target selecting u, $u_0 \leqq u \leqq u(t)$, is searched at t. If the condition

$$\Theta(t) > \frac{u_0^2}{4}\{(\frac{u_1}{u_0})^2 - \log(\frac{u_1}{u_0})^2 - 1\}$$

is satisfied, we define $u(t) = u_1$ irrespective of the solution to Eq. (2.42). Then, since the search region is given by

$$\mathbf{S}(t) = \{(r, \theta) \mid u_0 t \leqq r \leqq u(t) t\},$$

the R-optimal allocation is obtained by Eq.(2.39) as

$$\phi^*((r, \theta), t) = \frac{\Phi(t)}{\pi (u(t)^2 - u_0^2) t^2} \quad \text{for } u_0 t \leqq r \leqq u(t) t,$$

$$\phi^*((r, \theta), t) = 0 \quad \text{otherwise.}$$

The optimal stopping time T^* is determined by

$$\{2u(T^*)u_1 - u(T^*)^2 - u_0^2\}T^{*2} = \frac{R}{\pi c_0}.$$

2. Triangular Distribution

The speed distribution $g(u)$ is assumed to be the triangular distribution given by

(2.43)
$$g(u) = \frac{2u}{u_1^2} \quad \text{for} \quad 0 \leqq u \leqq u_1,$$
$$g(u) = 0 \quad \text{otherwise.}$$

From Eq.(2.40), the search should not be started if

$$\frac{R}{\pi c_0 u_1^2 t_0^2} < 1.$$

If the inequality sign is reversed, the R-optimal allocation $\boldsymbol{\phi}^*$ is obtained from Eq.(2.39).

(2.44)
$$\phi^*((r, \theta), t) = \frac{\Phi(t)}{\pi u_1^2 t^2} \quad \text{for} \quad 0 \leqq r \leqq u_1 t,$$
$$\phi^*((r, \theta), t) = 0 \quad \text{otherwise.}$$

The optimal stopping time T^* is calculated by Eq.(2.41) as

$$T^* = \{\frac{R}{\pi c_0 u_1^2}\}^{1/2}.$$

3. Rayleigh Distribution

Suppose the target selects its speed from the Rayleigh distribution;

$$g(u) = \frac{u}{\sigma^2} \exp(-\frac{u^2}{2\sigma^2}) \quad \text{for} \quad u \geqq 0.$$

If the system parameters satisfy the following condition,

$$\frac{R}{2\pi \sigma^2 c_0 t_0^2} < 1,$$

the search should not be started.

If the search should begin, the R-optimal allocation of searching effort $\boldsymbol{\phi}^*$ is

$$\phi^*((r, \theta), t) = \frac{\Phi(t)}{2\pi \sigma t^2 (2\Theta(t))^{1/2}} \quad \text{for } 0 \leqq r \leqq (8\sigma^2 \Theta(t))^{1/4} t,$$
$$\phi^*((r, \theta), t) = 0 \quad \text{otherwise.}$$

The optimal stopping time T^* is determined by

$$\{\sigma + (2\Theta(T^*))^{1/2}\} T^{*2} = \frac{R}{2\pi c_0 \sigma}.$$

4. Gamma Distribution

Let $g(u)$ be the gamma distribution given by

$$g(u) = \frac{\lambda^a u^{a-1} \exp(-\lambda u)}{\Gamma(a)} \quad \text{for} \quad u \geqq 0,$$

where a and λ are positive.

If the system parameters satisfy the relation;

$$\frac{\lambda^2}{2\pi \Gamma(a)} \left(\frac{a-2}{e}\right)^{a-2} \frac{R}{c_0 t_0^2} < 1 \quad \text{and} \quad a > 2,$$

the search should not begin.

If the search should be started, $\boldsymbol{\phi}^*$ is obtained as follows: If $0 < a \leqq 2$,

$$\phi^*((r, \theta), t) = \frac{\Phi(t)}{\pi\, u_1(t)^2 t^2} \qquad \text{for} \quad 0 \leqq r \leqq u_1(t)t,$$

$$\phi^*((r, \theta), t) = 0 \qquad \text{otherwise},$$

where $u_1(t)$ is the solution to the following equation

$$u_1(t)^3 + \frac{3(2-a)}{2\lambda} u_1(t)^2 - \frac{6}{\lambda}\Theta(t) = 0, \tag{2.45}$$

$$u_1(t) > 0.$$

This cubic equation is easily proved to have a positive root and two imaginary roots. If $a > 2$,

$$\phi^*((r, \theta), t) = \frac{\Phi(t)}{\pi\{u_1(t)^2 - u_0(t)^2\}t^2} \quad \text{for } u_0(t)t \leqq r \leqq u_1(t)t,$$

$$\phi^*((r, \theta), t) = 0 \qquad \text{otherwise},$$

where $u_0(t)$ and $u_1(t)$, $u_0(t) < u_1(t)$, are solutions to the following simultaneous equations:

$$\frac{a-2}{2} u_1(t)^2 \log \frac{u_1(t)}{u_0(t)} + \left(\frac{\lambda u_0(t)}{2} - \frac{a-2}{4}\right)(u_1(t)^2 - u_0(t)^2) - \frac{\lambda}{3}(u_1(t)^3 - u_0(t)^3) - \Theta(t) = 0, \tag{2.46}$$

$$(a-2)\log\frac{u_1(t)}{u_0(t)} - \lambda\,(u_1(t) - u_0(t)) = 0,$$

$$0 \leqq u_0(t) < u_1(t).$$

The optimal stopping time T^* is determined by the following equation. If $0 < a \leqq 2$,

$$\frac{R\delta_1(T^*)}{T^{*2}} = 2\pi c_0\{1 - \frac{\gamma(a, \lambda u_1(T^*))}{\Gamma(a)} + \frac{\delta_1(T^*)}{2} u_1(T^*)^2\},$$

where $u_1(t)$ is the solution to Eq. (2.45) and $\gamma(a, b)$ and $\delta_1(T)$ are defined by

$$\gamma(a, b) = \int_0^b x^{a-1}\exp(-x)\,dx, \tag{2.47}$$

$$\delta_i(T) = \frac{\lambda^a u_i(T)^{a-2}\exp(-\lambda u_i(T))}{\Gamma(a)}, \qquad i = 0, 1.$$

If $a > 2$,

$$\frac{R\delta_0(T^*)}{T^{*2}} = 2\pi c_0\{1 - \frac{\gamma(a, \lambda u_1(T^*)) - \gamma(a, \lambda u_0(T^*))}{\Gamma(a)} + \frac{\delta_0(T^*)}{2}(u_1(T^*)^2 - u_0(T^*)^2)\},$$

where $u_0(T^*)$ and $u_1(T^*)$ are the solutions to Eq. (2.46) at $t = T^*$ and $\gamma(a, b)$ and $\delta_0(T^*)$ are given by Eq. (2.47).

<u>Remark 2.8.</u> Theorem 2.9 states that the R-optimal allocation for

the diffusive F.C.D.M. target is identical to the uniformly P-optimal allocation. However, this property strongly depends on the assumptions of the model, particularly the homogeneity of the target space. To confirm this, we will consider a search situation in which the target space is inhomogeneous with respect to the reward. Suppose the reward depends on both the time and position where the target is detected, $R(y, t)$. In this case, since the P-optimal allocation does not depend on $R(y, t)$, the necessary and sufficient condition for the uniformly P-optimal allocation of the modified model is also given by Lemma 2.8. As mentioned before, the uniformly P-optimal allocation $\phi_T^0(y, t)$ for the F.C.D.M. target is distributed uniformly in the region $\mathbf{S}(t)$ given by Eq.(2.38). On the other hand, the necessary condition for the R-optimal allocation given by Theorem 2.3 is rewritten as:

$$\text{if } \phi^*(y, t) >(=) 0, \qquad -A_{yt}{}^T(\boldsymbol{\phi}_T^*) =(\leq) c_0(t)\lambda(t),$$

where

$$A_{yt}{}^T(\boldsymbol{\phi}_T^*) = \int_t^T \{R'(\eta_s(\eta_t^{-1}(y)), s) - c_0(s)\Phi(s)\} D_{yt}{}^s(\boldsymbol{\phi}_T^*)\,ds$$
$$+ c_0(t)Q_t(\boldsymbol{\phi}_T^*) - R(\eta_T(\eta_t^{-1}(y)), T) D_{yt}{}^T(\boldsymbol{\phi}_T^*),$$

$$D_{yt}{}^T(\boldsymbol{\phi}_T^*) = E_{yt}[b'(x, \int_{t0}^T \phi_T^*(\eta_s(x), s)\,ds)]\,q_t(y),$$

$$Q_T(\boldsymbol{\phi}_T^*) = 1 - \int_X p(x)\, b(x, \int_{t0}^T \phi_T^*(\eta_s(x), s)\,ds)\,dx.$$

However, in this case, the uniformly P-optimal allocation $\boldsymbol{\phi}_T^0$ is not identical to the R-optimal allocation $\boldsymbol{\phi}_T^*$. To prove this, we assume $\boldsymbol{\phi}_T^0 = \boldsymbol{\phi}_T^*$ and consider $(y, t) \in \mathbf{S}(t)$. Then, since Lemma 2.8 holds for $\boldsymbol{\phi}_T^*$, we substitute Eqs.(2.26a) and (2.26b) into $A_{yt}{}^T(\boldsymbol{\phi}_T^*)$ and we have

$$A_{yt}{}^T(\boldsymbol{\phi}_T^*) = \int_t^T \{R'(\eta_s(\eta_t^{-1}(y)), s) - c_0(s)\Phi(s)\} \frac{\mu(s)}{n(t)}\,ds$$
$$+ c_0(t)Q_t(\boldsymbol{\phi}_T^*) - R(\eta_s(\eta_t^{-1}(y)), T) \frac{\mu(T)}{n(t)}.$$

Since $\boldsymbol{\phi}_T^0 = \boldsymbol{\phi}_T^*$ and $(y, t) \in \mathbf{S}(t)$, $A_{yt}{}^T(\boldsymbol{\phi}_T^*)$ obtained from $\boldsymbol{\phi}_T^*$ must be equal to all (y, t). However, the right-hand side is a function of (y, t) and it contradicts the assumption $\boldsymbol{\phi}_T^0 = \boldsymbol{\phi}_T^*$. Therefore, $\boldsymbol{\phi}_T^0$ is not identical to $\boldsymbol{\phi}_T^*$. This means that Theorem 2.9 is valid only when the target space is homogeneous.

Remark 2.9. Danskin (1968) investigated a hide-and-search model for a diffusive F.C.D.M. target from the standpoint of game theory. In his model, the target is assumed to be a submarine contacted at $t=0$ and the searcher is helicopters on an antisubmarine mission. He deals with the problem with a discrete time space, a quantized searching effort (number of total glimpses by the helicopters' dipping sonar), and the detection function of cookie-cutter type. By employing the detection probability with n glimpses as the payoff function of the game, he formulates the problem as a two-person zero-sum game played on the velocity vector space of the target called a speed circle. He shows that the optimal strategy of the target is the triangular speed distribution given by Eq.(2.43). The concept of the speed circle corresponds to the parameter space **X** characterizing the target motion in our model. When the distribution of the target's speed is the triangular distribution, the probability density function of the target at t becomes uniform over the circle with the radius $u_1 t$, and therefore, it gives the optimal strategy for the target since the entropy of the target position is maximized. Although a few differences are caused by the divisibility of the searching effort, the optimal strategy of the searcher obtained by Danskin is conceptually identical with the allocation $\boldsymbol{\phi}^*$ given by Eq.(2.44).

Iida and Hiramoto (1984) investigated an optimal search problem for a target diffusing out from a datum point and obtained a suboptimal allocation of searching effort that maximizes the detection probability under the restriction of the searching effort rate. They deal with two types of moving targets, the diffusive F.C.D.M. target investigated in this section and random walk target. They formulate the moving target problem as an equivalent search problem for a stationary target by considering a transformation of the target space and the searching effort rate at each time. The P-optimal allocation and uniformly P-optimal allocation are obtained by applying the theory of the optimal search for the stationary target. This transformation at t is the distance reduced by $1/t$ to the F.C.D.M. target or by $1/\sqrt{t}$ to the random walk target, respectively. This transformation of the F.C.D.M. target implies that the target distribution on the two-dimensional target space is converted to the distribution on the velocity vector space of the target and, as mentioned earlier, it corresponds to the transformation of the target space **Y** to the motion parameter space **X** in our model. Therefore, the uniformly P-optimal allocation of the model for the diffusive F.C.D.M. target is completely identical with that of our model.

CHAPTER 3

Optimal Whereabouts Search Plan Minimizing the Expected Risk

3.1. Introduction

Imagine a search-and-rescue situation in which it is known that a lost party in a mountain can survive no longer than T units of time in the environment. If the searcher finds the party in some subarea up to T, the rescue can be effected immediately. If the party is not found until T, the rescue effort is assigned to some subarea; really it is a bet. The rescue will be successful if the lost party is in fact in the subarea. In analyzing this and similar situations, we will consider the following search model.

The target space consists of n boxes and a searcher wishes to localize a target within one of n boxes. The searcher may succeed either by detecting the target in his search or by guessing a box containing the target at the end of an unsuccessful search. This search-and-guess process is called whereabouts search.

The concept of whereabouts search was first introduced into search theory by Mela (1961). He discussed the measure of effectiveness for a search plan citing three criteria, i.e., the detection probability, expected information gain, and probability of correct commitment. The correct commitment probability is the probability of correctly stating the target's box either by detecting the target in the search or by guessing a box after an unsuccessful search. He considered special search problems in which a target is equally likely to be hidden in one of two or three boxes and the number of available looks is equal to the number of boxes, the conditional detection probability for the target is 0.5 in each box. By evaluating all alternatives of a search plan, he showed that the optimal plan maximizing the probability of correct commitment is to search all the boxes but one that the searcher intends to guess if he fails to detect the target. Tognetti (1968) named this search plan "the miss one strategy" and the probability of correct commitment "the whereabouts probability." Tognetti generalized Mela's model assuming a uniform distribution of the target in n boxes and arbitrary value of the conditional detection probability α. He proved that the miss one strategy is optimal for the whereabouts search model for any values of n and α. Kadane (1971) investigated the whereabouts search model in a general form and proved theorems. He considered an n-box problem with a general target distribution in which the conditional detection probability and the search cost depend not only on the box to be searched but also on the number of looks assigned to the box previously. A definite amount of total searching cost C is assumed to be available to the searcher and the problem is formulated as maximizing the whereabouts probability subject to the restriction of the total searching cost. He showed that the miss one strategy is optimal generally in the whereabouts search maximizing the whereabouts probability, some box should be chosen to be guessed after an unsuccessful search. The optimal whereabouts search plan for all other boxes coincides with the optimal detection search plan. He also derived some sufficient conditions for the optimal box to be guessed and proposed a branch-and-bound algorithm to find the whereabouts search plan. Stone and Kadane (1981) investigated a whereabouts search plan maximizing the whereabouts probability for a moving target. By applying the theorems of the optimal whereabouts search for a stationary target given by Kadane (1971), they showed that solving the optimal whereabouts search problem for a moving target results in solving a finite number of the optimal detection search problem for the moving target. Iida and Kanbashi (1983) investigated an optimal whereabouts search plan that minimizes the expected risk for a stationary target, and Iida (1988a) studied the optimal whereabouts search plan for a mortal target.

In this chapter, we will investigate an optimal search-and-guess plan that minimizes the expected risk of the whereabouts search process. In the next section, we will deal with the whereabouts search for a stationary target and derive a necessary and sufficient condition for the optimal allocation of the searching effort and necessary conditions for the optimal stopping time of the search and the optimal box to be guessed. In Section 3.3, the optimal whereabouts search plan for a mortal target is investigated.

3.2. Optimal Whereabouts Search Plan for an Appeared Target

3.2.1. Optimal Distribution of Searching Effort

In this section, we will study the optimal whereabouts search plan for an appeared, stationary and immortal target minimizing the expected risk. The results of this section are based on Iida and Kanbashi (1983).

The target space is assumed to consist of n boxes and a target is known to be in one of the boxes with a probability vector $\mathbf{p} = \{p_i\}$, where p_i is the initial probability of the target being in box i, $p_i > 0$, $i = 1, 2, \cdots, n$, $\Sigma_i p_i = 1$. We assume that the search in box i costs c_i (>0) per unit density of searching effort and a definite total searching cost C per unit time is available to the searcher. It is assumed that the search is carried out continuously within the limited total searching cost C per unit time, in which C can be divided arbitrarily and be allocated to any boxes. The conditional detection probability $b(i, \psi)$, given that the target is hidden in box i and density ψ of searching effort is applied to the box, is assumed to be an exponential function of ψ, $b(i, \psi) = 1 - \exp(-\alpha_i \psi)$, which corresponds to the detection law of the random search. The searcher gains a reward R_i (≧0) when he successfully detects the target in box i. Meanwhile, if the search is continued without any detection, the searcher can stop searching whenever he wants by guessing a box. It is assumed that a cost W_i (≧0) is imposed on the searcher when he stops the search by guessing box i, and if the target is contained in the box (called a correct guess), he earns a reward R_i^0 (≧0). A decision triplet how to search, when to guess, and which box to be guessed is now named a whereabouts search plan. The whereabouts search plan is denoted by $(\boldsymbol{\phi}, T, k)$ in which $\boldsymbol{\phi} = \{\phi(i, t), i = 1, 2, \cdots, n, t \in [0, T]\}$ is a distribution of searching effort density in $[0, T]$, T is the stopping time of the search, and k is the box to be guessed. Let $\psi = \{\psi(i, t), i = 1, 2, \cdots, n, t \in [0, T]\}$ be the set of the cumulative searching effort density allocated to box i up to t. Then, the relation $\phi(i, t) = \partial \psi(i, t)/\partial t$ holds for a.e. $t \in [0, T]$. The derivative $\partial \psi(i, t)/\partial t$ at the external point $t = 0$ or T is defined by the right or left derivative, respectively.

As the measure of effectiveness for the whereabouts search plan, we adopt the expected risk, namely, the expected cost imposed by searching and guessing minus the expected reward obtained by detection of the target or guessing the box containing the target correctly. The expected risk $f(\boldsymbol{\phi}, T, k)$ of a whereabouts search plan $(\boldsymbol{\phi}, T, k)$ is given by

$$f(\boldsymbol{\phi}, T, k) = \Sigma_i p_i [\int_0^T \alpha_i \phi(i, t) \{\Sigma_j c_j \psi(j, t) - R_i\} \exp(-\alpha_i \psi(i, t)) dt$$
$$+ \exp(-\alpha_i \psi(i, T)) \{\Sigma_j c_j \psi(j, T) + W_k\}] - p_k \exp(-\alpha_k \psi(k, T)) R_k^0.$$

Since $\phi(i, t)$ is searching effort density allocated to box i at t, a constraint of the total cost rate, $\Sigma_i c_i \phi(i, t) \leq C$ is imposed for all $t \in [0, T]$. However, since the target distribution, cost, and reward are stationary in our problem, the optimal search uses all the searching cost rate C exhaustively each time. Therefore, the inequality in the constraint can be replaced by an equality and the term $\Sigma_i c_i \psi(i, t)$ in the expression is equal to Ct. Substituting this relation into the equation and integrating by parts, we have the following simplified expression,

$$(3.1)\qquad f(\boldsymbol{\phi}, T, k) = \Sigma_i p_i[C\int_0^T \exp(-\alpha_i \varphi(i,t))dt - R_i(1-\exp(-\alpha_i \varphi(i,T))) + (W_k - \delta_{ik}R_k^0)\exp(-\alpha_i \varphi(i,T))].$$

The following restrictions are imposed.

$$(3.2)\qquad \begin{aligned} &\Sigma_i c_i \phi(i,t) = C \quad \text{for all } t \in [0,T], \\ &\phi(i,t) \geqq 0 \quad \text{for all } i \text{ and } t \in [0,T], \\ &T \geqq 0. \end{aligned}$$

Hence, our problem is formulated as a variational problem to find an optimal whereabouts search plan $(\boldsymbol{\phi}, T, k)$ that minimizes the expected risk $f(\boldsymbol{\phi}, T, k)$ given by Eq.(3.1) subject to the restrictions given by Eq.(3.2). The conditions for the plan $(\boldsymbol{\phi}, T, k)$ to be optimal are derived by three steps. First, we deal with the minimization problem of Eq.(3.1) by conditioning both a stopping time T of search and a box k to be guessed and derive a conditionally optimal allocation of searching effort density. This allocation is called the conditionally optimal search plan, and is denoted by $\boldsymbol{\phi}_{Tk}{}^* = \{\phi_{Tk}{}^*(i,t), i=1,2,\cdot\cdot,n, t \in [0,T]\}$. Next, by conditioning the box k to be guessed and the conditionally optimal search plan $\boldsymbol{\phi}_{Tk}{}^*$ until T, we choose the stopping time T so as to minimize $f(\boldsymbol{\phi}_{Tk}{}^*, T, k)$. We denote this conditionally optimal stopping time as $T_k{}^*$. Finally, by minimizing $f(\boldsymbol{\phi}_{Tk}{}^*, T_k{}^*, k)$ with respect to k, we determine the optimal box k^* to be guessed. Then the triplet $(\boldsymbol{\phi}_{Tk*}{}^*, T_{k*}{}^*, k^*)$ is the optimal whereabouts search plan. From now on, the subscripts T_k and k in $\boldsymbol{\phi}_{Tk}{}^*$ and $T_k{}^*$ are omitted if there is no confusion.

<u>Theorem 3.1.</u> *A necessary condition for the conditionally optimal search plan is that there exists a nonnegative function $\lambda(t)$ such that if $\phi^*(i,t) > 0$,*

$$(3.3a)\qquad \frac{\alpha_i p_i}{c_i}[R_{ik}\exp(-\alpha_i \varphi^*(i,T)) + C\int_t^T \exp(-\alpha_i \varphi^*(i,z))dz] = \lambda(t),$$

if $\phi^*(i,t) = 0$,

$$(3.3b)\qquad \frac{\alpha_i p_i}{c_i}[R_{ik}\exp(-\alpha_i \varphi^*(i,T)) + C\int_t^T \exp(-\alpha_i \varphi^*(i,z))dz] \leqq \lambda(t),$$

for all i and a.e. t in $[0,T]$, where R_{ik} is defined by

$$(3.4)\qquad \begin{aligned} &R_{ik} = R_i + W_k - \delta_{ik}R_k^0, \\ &\delta_{ik} = 1 \text{ if } i = k, \text{ and } \delta_{ik} = 0 \text{ if } i \neq k. \end{aligned}$$

If $R_{ij} \geqq 0$, these conditions are also sufficient for $\boldsymbol{\phi}^*$ to be *conditionally optimal.*

Proof. By applying Lemma 2.2, we can prove the theorem. The functional $f(\boldsymbol{\phi}, T, k)$ defined by Eq. (3.1) has a Gateaux differential at $\boldsymbol{\phi}$ to any direction e in $\boldsymbol{\phi}+\varepsilon e \in \boldsymbol{\Psi}_0$ for $\varepsilon \ll 1$. From the definition of the Gateaux differential given by Eq. (2.10), we obtain

$$\delta f(\boldsymbol{\phi}, T, k, e) = -\Sigma_i p_i \alpha_i [R_{ik} \exp(-\alpha_i \varphi(i, T)) \int_0^T e(i, t) dt$$
$$+ \int_0^T \{C \exp(-\alpha_i \varphi(i, t)) \int_0^t e(i, s) ds\} dt].$$

By interchanging the order of the integrals in the braces, we have

$$\delta f(\boldsymbol{\phi}, T, k, e) = -\Sigma_i \int_0^T e(i, s) p_i \alpha_i [R_{ik} \exp(-\alpha_i \varphi(i, T))$$
$$+ \int_t^T C \exp(-\alpha_i \varphi(i, z)) dz] ds.$$

Therefore, from the definition of the kernel of the Gateaux differential given by Eq. (2.11), the kernel of $\delta f(\boldsymbol{\phi}, T, k, e)$ is calculated as

$$d(\boldsymbol{\phi}, T, k) = -p_i \alpha_i [R_{ik} \exp(-\alpha_i \varphi(i, T)) + \int_t^T C \exp(-\alpha_i \varphi(i, z)) dz].$$

Here if we set $F(\boldsymbol{\phi}) = -f(\boldsymbol{\phi}, T, k)$, Lemma 2.2 can be applied to our problem and the necessary conditions given by Eqs. (3.3a) and (3.3b) are derived by replacing $c(i, t)$ with c_i in the lemma.

The sufficiency of the conditions in Eqs. (3.3a) and (3.3b) are also proved by using Lemma 2.2. The convexity of the expected risk $f(\boldsymbol{\phi}, T, k)$ with respect to $\boldsymbol{\phi}$ is proved as follows. Let $\boldsymbol{\phi}^1$ and $\boldsymbol{\phi}^2$ be arbitrary effort allocations that satisfy the constraints in Eq. (3.2) and set $\boldsymbol{\phi} = (1-\theta)\boldsymbol{\phi}^1 + \theta \boldsymbol{\phi}^2$, $0 \leq \theta \leq 1$. Then we have

$$f(\boldsymbol{\phi}, T, k) = \Sigma_i p_i [\int_0^T C \exp\{-\alpha_i((1-\theta)\varphi^1(i, t) + \theta \varphi^2(i, t))\} dt$$
$$- R_i + R_{ik} \exp\{-\alpha_i((1-\theta)\varphi^1(i, T) + \theta \varphi^2(i, T))\}].$$

Since the negative exponential function is convex, we have

$$\exp\{-\alpha_i((1-\theta)\varphi^1(i, t) + \theta \varphi^2(i, t))\}$$
$$\leq (1-\theta)\exp(-\alpha_i \varphi^1(i, t)) + \theta \exp(-\alpha_i \varphi^2(i, t)).$$

Therefore, we have the following relation, if $R_{ik} \geq 0$,

$$f(\boldsymbol{\phi}, T, k) \leq \Sigma_i p_i [\int_0^T C \{(1-\theta)\exp(-\alpha_i \varphi^1(i, t))$$
$$+ \theta \exp(-\alpha_i \varphi^2(i, t))\} dt - R_i + R_{ik}\{(1-\theta)\exp(-\alpha_i \varphi^1(i, T))$$
$$+ \theta \exp(-\alpha_i \varphi^2(i, T))\}]$$
$$= (1-\theta) f(\boldsymbol{\phi}^1, T, k) + \theta f(\boldsymbol{\phi}^2, T, k).$$

The equality holds only for $\theta = 0$ or $\theta = 1$, therefore, $f(\boldsymbol{\phi}, T, k)$ is strictly convex with respect to $\boldsymbol{\phi}$. In our model, $\boldsymbol{\phi}$ is limited to $\boldsymbol{\Psi}_0$ as described before. Therefore, the necessary conditions given by Eqs. (3.3a) and (3.3b) are also sufficient for $\boldsymbol{\phi}^*$ to be optimal by Lemma 2.2.

□

Remark 3.1. An interpretation of Theorem 3.1 is as follows: The cumulative density of searching effort allocated to box i from t to s is denoted by $\zeta(i,t,s) = \int_t^s \phi(i,z)dz$. Then using this relation, we can rewrite Eqs.(3.3a) and (3.3b) as: If $\phi^*(i,t) >(=) 0$,

$$\frac{\alpha_i p_i \exp(-\alpha_i \phi^*(i,t))}{c_i}[R_i + \{C\int_t^T \exp(-\alpha_i \zeta^*(i,t,z))dz$$

$$- R_i(1-\exp(-\alpha_i \zeta^*(i,t,T))) + (W_k - \delta_{ik}R_k^0)\exp(-\alpha_i \zeta^*(i,t,T))\}] \tag{3.5}$$

$$=(\leqq) \lambda(t).$$

Each term of this relation is interpreted as follows. The numerator of the first term, $\alpha_i p_i \exp(-\alpha_i \phi^*(i,t))$, represents the probability of detecting the target with unit density of searching effort allocated to box i at t. The first term in the brackets is the reward gained by detecting the target in box i. By comparing Eq.(3.5) with Eq.(3.1), we see that the term in the braces is the expected risk of the whereabouts search in $[t,T]$ given that the target is in box i and is not yet detected at t. Since the search is to be continued until T if the target is not detected and then box k is to be guessed, the quantity of this term is interpreted as the saved risk earned by detecting the target in box i at t. Hence, if we detect the target in box i at t, we gain the direct reward and save the expected risk that is to be incurred after t. Thus the quantity in the brackets multiplied by the numerator of the first term represents the expected marginal return. Therefore, the left-hand side of Eq.(3.5) is interpreted as the ratio of the expected marginal return versus searching cost in allocation of unit density of searching effort in box i at t. Hence, Theorem 3.1 can be paraphrased as follows: If searching effort is to be allocated to box i at t, the density of the searching effort should be determined in such a way that the expected marginal return versus cost ratio is balanced to $\lambda(t)$ in the boxes being searched at t, and if box i should not be searched, this is because box i does not have a larger expected marginal return versus cost ratio than $\lambda(t)$.

It should be noted that if $R_{ik} = R_i$ for all i and k, Theorem 3.1 becomes identical to Theorem 1.1. This correspondence is not surprising, because if $R_{ik} = R_i$ then $W_k - \delta_{ik}R_k^0 = 0$ from Eq.(3.4) and the objective function $f(\boldsymbol{\phi},T,k)$ given by Eq.(3.1) is identical with $f(\boldsymbol{\phi}_T)$ given by Eq.(1.1). The relations $W_k - \delta_{ik}R_k^0 = 0$ for all i and k mean $W_k = R_k^0 = 0$, and in this case, guessing a box is nonsense and the whereabouts search model becomes a detection search model.

From now on, we assume $R_{ik} > 0$. Since R_{ik} is not dependent on $\boldsymbol{\phi}$ as shown in Eq.(3.4), corollaries similar to Corollaries 1.2 and 1.3 and Theorems 1.4 and 1.5 derived from Theorem 1.1 are also established from Theorem 3.1. Since the derivation is in exactly the same pattern, the corollaries are presented without proofs.

<u>Corollary 3.2.</u> *If* $\phi^*(i,T) > 0$,

$$\frac{\alpha_i p_i}{c_i} R_{ik} \exp(-\alpha_i \phi^*(i,T)) = \lambda(T), \tag{3.6a}$$

if $\phi^*(i,T) = 0$,

$$\frac{\alpha_i p_i}{c_i} R_{ik} \exp(-\alpha_i \phi^*(i,T)) \leqq \lambda(T). \tag{3.6b}$$

<u>Corollary 3.3.</u> *The function* $\lambda(t)$ *in Theorem* 3.1 *is a continuous, strictly decreasing and strictly convex function of t.*

<u>Theorem 3.4.</u> *Suppose* $\phi^*(i,t)$ *is not dense at* t_1 *and* t_2, $0 \leqq t_1 < t_2 \leqq T$.

1. *If* $\phi^*(i,t)$ *is positive in the interval* $[t_1, t_2)$ *and is zero at* t_2, $\phi^*(i,t)$ *is always zero throughout the interval* $[t_2, T]$.
2. *If* $\phi^*(i,t)$ *is positive at both* t_1 *and* t_2, *then* $\phi^*(i,t)$ *is always positive in the interval* $[t_1, t_2]$.
3. *If both* $\phi^*(i,t)$ *and* $\phi^*(j,t)$ *are positive in the interval* $[t_1, t_2]$, *then at any time t in* $[t_1, t_2]$,

$$\frac{\alpha_i p_i}{c_i} \exp(-\alpha_i \phi^*(i,t)) = \frac{\alpha_j p_j}{c_j} \exp(-\alpha_j \phi^*(j,t)), \tag{3.7}$$

$$\alpha_i \phi^*(i,t) = \alpha_j \phi^*(j,t) = \frac{C}{\Sigma_{S(t)} (c_i/\alpha_i)},$$

where $\mathbf{S}(t) = \{i \mid \phi^*(i,t) > 0\}$.

<u>Theorem 3.5.</u> *Suppose both boxes i and j are boxes searched until T. Let* T_i^{s*} *and* T_i^{e*} *be the optimal starting time and the stopping time of the search in box i.*

1. *If and only if* $\alpha_i p_i / c_i \geqq \alpha_j p_j / c_j$, *then* $T_i^{s*} \leqq T_j^{s*}$.
2. *If and only if*

$$\frac{\alpha_i p_i}{c_i} \exp(-\alpha_i \phi^*(i,T)) \leqq \frac{\alpha_j p_j}{c_j} \exp(-\alpha_j \phi^*(j,T)),$$

then $T_i^{e*} \geqq T_j^{e*}$.

3. *If and only if* $R_{ik} \geqq R_{jk}$, *then* $T_i^{e*} \geqq T_j^{e*}$.

3.2.2. Optimal Stopping Time of the Search

In the previous section, the stopping time T was assumed to be given. In this section, however, T is considered a variable and the stopping time that minimizes the expected risk given box k to be guessed is sought. We shall call it the conditionally optimal stopping time for box k and denote it T_k^*. As to the effort allocation associated with T_k, we will consider only the conditionally optimal allocation of searching effort obtained in the previous section.

Theorem 3.6. *A necessary condition for the conditionally optimal stopping time T_k^* of the search given box k to be guessed is*

$$\lambda(T_k^*-\varepsilon) \geq Q(T_k^*-\varepsilon),$$
$$\lambda(T_k^*) = Q(T_k^*),$$
$$\lambda(T_k^*+\varepsilon) \leqq Q(T_k^*+\varepsilon),$$

where ε is a sufficiently small time interval and $\lambda(T)$ is given by Eqs.(3.6a) and (3.6b). $Q(T)$ is the nondetection probability of the target with $\boldsymbol{\phi}_{Tk}^*$ up to T_k given by

$$Q(T) = \Sigma_i p_i \exp(-\alpha_i \phi_{Tk}^*(i,T)).$$

This theorem corresponds to Theorem 1.7. In Theorem 1.7, if R_i in $\lambda(T)$ given by Corollary 1.2 is replaced by R_{ik}, we obtain Theorem 3.6. Therefore the proof of Theorem 3.6 is identical with that of Theorem 1.7 since R_{ik} does not depend on $\boldsymbol{\phi}_{Tk}^*$. Furthermore, all the corollaries derived from Theorem 1.7, are derived similarly from Theorem 3.6 by replacing R_i with R_{ik}. We show these corollaries without proofs.

Corollary 3.7. *A necessary condition for the conditionally optimal stopping time T_k^* given a box k to be guessed is*

$$\frac{\alpha_i p_i(T) R_{ik}}{c_i} \geqq 1 \qquad if \quad T < T_k^*,$$
$$\frac{\alpha_i p_i(T) R_{ik}}{c_i} = 1 \qquad if \quad T = T_k^*,$$
$$\frac{\alpha_i p_i(T) R_{ik}}{c_i} \leqq 1 \qquad if \quad T > T_k^*,$$

for any T in the neighborhood of T_k^* and for any $i \in \mathbf{S}(T_k^*)$.

Corollary 3.8. 1. *A necessary condition for $T_k^* = 0$ is*

$$\max_i \frac{\alpha_i p_i R_{ik}}{c_i} \leqq 1.$$

2. *If $\max_i \alpha_i p_i(T) R_{ik}/c_i > 1$, the search should not be stopped at T.*
3. *If $R_{ik} \leqq c_i/\alpha_i$ for all i, then $T_k^* = 0$.*

Theorem 3.9. 1. *A necessary condition for $T_k^* = \infty$ is*

$$\{\Sigma_i \frac{c_i}{\alpha_i}\}^2 \leqq \Sigma_i \frac{c_i R_{ik}}{\alpha_i}.$$

2. *A sufficient condition for $T_k^* = \infty$ is*

$$\Sigma_i \frac{c_i}{\alpha_i R_{ik}} \leqq 1.$$

Theorem 3.10. *Suppose* $0 < T_k^* < \infty$. *Let* $\mathbf{S}(t)$ *and* $\mathbf{D}(t)$ *be the sets defined by Eqs.* (1.6a) *and* (1.6b).

1. *If* $\alpha_i p_i R_{ik}/c_i > 1$, *box i is an element of* $\mathbf{D}(T_k^*)$.
2. *If* $i \in \mathbf{S}(T_k^*)$ *and* $\alpha_i p_i R_{ik}/c_i < \alpha_j p_j R_{jk}/c_j$, *then box j is an element of* $\mathbf{D}(T_k^*)$.
3. *If* $i \in \mathbf{D}(T_k^*)$ *and* $\alpha_i p_i R_{ik}/c_i < \alpha_j p_j R_{jk}/c_j$ *and* $\alpha_i p_i/c_i < \alpha_j p_j/c_j$, *box j is also an element of* $\mathbf{D}(T_k^*)$.
4. *If* $i \in \mathbf{S}(T_k^*)$, R_{ik} *is larger than* $\Sigma_{S(T_k^*)} c_i/\alpha_i$.

Remark 3.2. The meaning of the condition for the conditionally optimal stopping time T_k^* is elucidated as follows. As seen by Corollary 3.7, the quantity, $\alpha_i p_i(T_k^*)R_{ik}/c_i$ for box $i \in \mathbf{S}(T_k^*)$, has an important role for T_k^*. In this quantity, $\alpha_i p_i(T_k^*)$ is the marginal detection probability in box i at T_k^* and R_{ik} is the sum of the direct reward and the saved risk earned by detecting the target in box i at T_k^*. On the other hand, the expected marginal return versus searching cost ratio has the maximum value for box $i \in \mathbf{S}(T_k^*)$ from Corollary 3.2. Therefore, Corollary 3.7 is interpreted as follows: The largest value of the expected marginal return versus searching cost ratio decreases across unity as the stopping time increases across the conditionally optimal stopping time. Box k should be guessed at the instant when the largest expected marginal return $\alpha_i p_i(T_k^*)R_{ik}$ becomes as small as the cost c_i.

Remark 3.3. As described in the first section of this chapter, the optimal whereabouts search plan given by Kadane (1971) that maximizes the whereabouts probability is to search with the effort distribution determined by balancing the marginal detection probability among boxes except the box to be guessed, and the box to be guessed should not be searched at all; $\varphi^*(k^*, T_k^*) = 0$. This search plan (the miss one strategy) is not optimal for our whereabouts search model. In our model, if $\alpha_k p_k(R_k + W_k - R_k^0)/c_k > 1$, then $\varphi^*(k, T_k) > 0$. This statement is proved by the reductive absurdity. We assume $\varphi^*(k, T_k) = 0$ under this condition. Since $\alpha_k p_k R_{kk}/c_k > 1$, to guess immediately is not optimal from Corollary 3.8-2; i.e., $T_k^* > 0$. Let box j be an element of $\mathbf{S}(T_k^*)$. From Corollary 3.7, we obtain

$$\begin{aligned}\frac{\alpha_k p_k R_{kk}}{c_k} &\leq \frac{\alpha_j p_j R_{jk}}{c_j}\exp(-\alpha_j \varphi^*(j, T_j^*))\\ &= \Sigma_i p_i \exp(-\alpha_i \varphi^*(i, T_i^*)).\end{aligned}$$

This deduces a contradiction $\Sigma_i p_i \exp(-\alpha_i \varphi^*(i, T_i^*)) > 1$. Hence, we can conclude $\varphi^*(k, T_k^*) > 0$, namely, we should search the box to be guessed in this case.

3.2.3. Optimal Box to Be Guessed

In this section, employing the conditionally optimal allocation of searching effort density $\phi_{T_k}{}^*$ and the stopping time $T_k{}^*$ derived in the previous sections, we minimize the expected risk $f(\phi_{T_k}{}^*, T_k{}^*, k)$ by choosing a box to be guessed. Let k^* be the optimal box to be guessed. Then the triplet $(\phi_{T_{k*}}{}^*, T_{k*}{}^*, k^*)$ is the optimal whereabouts search plan for our problem.

<u>Theorem 3.11.</u> *A necessary condition for the optimal box k^* to be guessed is*

$$W_{k*} - p_{k*}(T_{k*}{}^*)R_k{}^0 \leqq W_j - p_j(T_{k*}{}^*)R_j{}^0$$

for any j, $j \neq k^*$.

<u>Proof.</u> Suppose the optimal whereabouts search plan is $(\phi_{T_{k*}}{}^*, T_{k*}{}^*, k^*)$. Here we consider a plan $(\phi_{T_{k*}}{}^*, T_{k*}{}^*, j)$ in which the allocation of effort density and the stopping time are the same as the optimal plan but the box to be guessed is interchanged with another box j. Then obviously

$$f(\phi_{T_{k*}}{}^*, T_{k*}{}^*, k^*) \leqq f(\phi_{T_{k*}}{}^*, T_{k*}{}^*, j).$$

From this, we have

$$f(\phi_{T_{k*}}{}^*, T_{k*}{}^*, k^*) - f(\phi_{T_{k*}}{}^*, T_{k*}{}^*, j)$$
$$= Q(T_{k*}{}^*)[W_{k*}-p_{k*}(T_{k*}{}^*)R_{k*}{}^0 - (W_j-p_j(T_{k*}{}^*)R_j{}^0)] \leqq 0.$$

Therefore, $W_{k*} - p_{k*}(T_{k*}{}^*)R_{k*}{}^0 \leqq W_j - p_j(T_{k*}{}^*)R_j{}^0$ for any j. □

<u>Theorem 3.12.</u> *Suppose $R_{ik} > 0$ for all i.*

1. *If $\Sigma_j(c_j/\alpha_j) < \min_i[R_i+W_k-R_k{}^0]$, box k is not the optimal box to be guessed.*

2. *If $\Sigma_i(c_i/\alpha_i R_{ik}) < 1$, box k is not the optimal box to be guessed.*

<u>Proof.</u> In order to prove the theorem by using the reductive absurdity, we assume that box k is the optimal box to be guessed and the optimal whereabouts search plan is given by $(\phi_{T_k}{}^*, T_k{}^*, k)$.

1. Let ϕ be an effort allocation such that $\phi = \phi_{T_k}{}^*$ in $[0, T_k{}^*]$ and $\phi = \{C/(\alpha_i \Sigma_j(c_j/\alpha_j))\}$ in $(T_k{}^*, \infty)$. Then we have

$$f(\phi, \infty, \cdot) - f(\phi_{T_k}{}^*, T_k{}^*, k)$$
$$= \Sigma_i p_i \exp(-\alpha_i \psi(i, T_k{}^*))[\Sigma_j \frac{c_j}{\alpha_j} - R_i - W_k + p_k(T_k{}^*)R_k{}^0]$$
$$\leqq \Sigma_i p_i \exp(-\alpha_i \psi(i, T_k{}^*))[\Sigma_j \frac{c_j}{\alpha_j} - R_i - W_k + R_k{}^0].$$

Since $\Sigma_j(c_j/\alpha_j) - R_i - W_k + R_k{}^0 < 0$ for any i by the condition of the theorem, $f(\phi, \infty, \cdot) < f(\phi_{T_k}{}^*, T_k{}^*, k)$ is deduced and it contradicts the assumption that the optimal whereabouts search plan is $(\phi_{T_k^*}, T_k^*, k)$.

2. If $(\phi_{T_k^*}, T_k^*, k)$ is the optimal plan, from Corollary 3.7, an inequality $\alpha_i p_i(T_k^*+\varepsilon)R_{ik}/c_i \leqq 1$ holds for any box i. Hence, $\Sigma_i(c_i/\alpha_i R_{ik}) \geqq \Sigma_i p_i(T_k^*+\varepsilon) = 1$ is derived, and this contradicts the assumption of the theorem. □

The following corollary is derived as the contrapositive statement of Theorem 3.12.

Corollary 3.13. *Necessary conditions for box k to be optimal are*

$$\Sigma_j \frac{c_j}{\alpha_j} \geqq \min_i[R_i + W_k - R_k^0],$$

$$\Sigma_j \frac{c_j}{\alpha_j R_{jk}} \geqq 1.$$

Remark 3.4. We consider the meaning of Theorem 3.11. Suppose the search is conducted by using the conditionally optimal allocation $\phi_{T_{k*}^*}$ and it fails to detect the target until T_{k*}^*. Since the expected risk of guessing box j at T_{k*}^* is $W_j - p_j(T_{k*}^*)R_j^0$, Theorem 3.11 is interpreted as follows: If box k is the optimal box to be guessed, box k has the smallest expected risk among boxes at T_{k*}^*.

3.3. Optimal Whereabouts Search Plan for a Mortal Target

In this section, we will study a whereabouts search model for a target with a random lifetime. This model was investigated by Iida (1988a).

A stationary target is known to be in one of n boxes and the probability p_i of the target being in box i is given to the searcher ($p_i>0$, $\Sigma_i p_i=1$). The lifetime of the target in box i is a random variable with a defective probability distribution function $F_i(t)$ that has a probability mass β_i at $t = 0$ (the probability that the target in box i has died before the start of the search). Let γ_i be the probability that the target in box i never dies. $F_i(t)$ is assumed to be differentiable with respect to t in $(0,\infty)$ and the derivative of $F_i(t)$ is denoted by $f_i(t)$. Therefore, we have

$$F_i(t) = \beta_i + \int_0^t f_i(s)\,ds,$$

$$\beta_i + \int_0^\infty f_i(s)\,ds + \gamma_i = 1.$$

We assume that the detection of the target is not influenced by the lifetime of the target; whether the target is alive or has died does not influence the detection law of the searcher. The searcher obtains a reward R_i when the target is detected alive in box i, but no reward is gained by the detection of the target if it is not alive. We also make a similar assumption for the correct guess; if the target is alive when the searcher guesses the box i correctly, the searcher gains a reward R_i^0, otherwise, no reward is gained by the searcher.

We assume that the whereabouts search process investigated in this section is identical to the model dealt with in the previous section except for the assumptions about the lifetime of the target and the rewards R_i and R_i^0. The expected risk of whereabouts search $f(\boldsymbol{\phi}, T, k)$ is given by

$$(3.8) \qquad f(\boldsymbol{\phi}, T, k) = \Sigma_i p_i [\int_0^T \alpha_i \phi(i, t)\{\Sigma_i c_i \varphi(i, t) - R_i(1-F_i(t))\} \times \exp(-\alpha_i \varphi(i, t))dt + \{\Sigma_i c_i \varphi(i, t) + W_k - \delta_{ik} R_k^0 (1-F_k(T))\}\exp(-\alpha_i \varphi(i, T))],$$

where $\delta_{ik} = 1$ if $i = k$, and $\delta_{ik} = 0$ if $i \neq k$.

By the assumption of the model, the total searching cost rate at t is restricted within C, $\Sigma_i c_i \phi(i, t) \leqq C$, for all $t \in [0, T]$. However, in our model, the target is assumed to die with the passing of time and to be stationary, and neither the reward nor the searching cost changes in time. Therefore, any allocation of searching effort in which the total cost rate C is not used up during the search is not optimal. Hence, we can omit the inequality sign in the restriction, $\Sigma_i c_i \phi(i, t) \leqq C$, without any influence on the optimal solution and we have $\Sigma_i c_i \varphi(i, t) = Ct$. Substituting this relation into Eq. (3.8) and integrating by parts, we obtain a simplified expression of the expected risk $f(\boldsymbol{\phi}, T, k)$.

$$(3.9) \qquad f(\boldsymbol{\phi}, T, k) = \Sigma_i p_i [R_{ik}(T)\exp(-\alpha_i \varphi(i, T)) - R_i(1-\beta_i) + \int_0^T (C + R_i f_i(t))\exp(-\alpha_i \varphi(i, t))dt],$$

$$R_{ik}(T) = (R_i - \delta_{ik} R_k^0)(1-F_k(T)) + W_k.$$

In addition to the constraint C, $\phi(i, t)$ must be nonnegative for all i and $t \in [0, T]$, and T is also nonnegative. Hence the following constraints are imposed.

$$(3.10) \qquad \begin{aligned} &\Sigma_i c_i \phi(i, t) = C \quad \text{for all } t \in [0, T], \\ &\phi(i, t) \geqq 0 \quad \text{for all } i \text{ and } t \in [0, T], \\ &T \geqq 0. \end{aligned}$$

Therefore, the problem is formulated as a variational problem to find an optimal whereabouts search plan $(\boldsymbol{\phi}, T, k)$ that minimizes the expected risk $f(\boldsymbol{\phi}, T, k)$ given by Eq. (3.9) subject to the constraints given by Eq. (3.10).

The conditionally optimal allocation of searching effort given both the stopping time T and the box k to be guessed are obtained by the following theorem.

Theorem 3.14. *Suppose $R_{ik}(T) \geqq 0$ for all i. A necessary and sufficient condition for a conditionally optimal allocation of searching effort $\boldsymbol{\phi}^*$ is that there exists a nonnegative function $\lambda(t)$ such that for a.e. t and i, if $\phi^*(i,t) > 0$,*

(3.11a)
$$\frac{\alpha_i p_i}{c_i}\Big[R_{ik}(T)\exp(-\alpha_i\phi^*(i,T)) + \int_t^T (C+R_i f_i(z))\exp(-\alpha_i\phi^*(i,z))\,dz\Big] = \lambda(t),$$

if $\phi^*(i,t) = 0$,

(3.11b)
$$\frac{\alpha_i p_i}{c_i}\Big[R_{ik}(T)\exp(-\alpha_i\phi^*(i,T)) + \int_t^T (C+R_i f_i(z))\exp(-\alpha_i\phi^*(i,z))\,dz\Big] \leqq \lambda(t).$$

Proof. The proof of the theorem is given by following a pattern of reasoning similar to that for Theorem 3.1. The Gateaux differential of the functional $f(\boldsymbol{\phi},T,k)$ is calculated as

$$\delta f(\boldsymbol{\phi},T,k,e) = -\Sigma_i p_i\alpha_i\Big[R_{ik}(T)\exp(-\alpha_i\phi(i,T))\int_0^T e(i,t)\,dt + \int_0^T\Big\{(C+R_i f_i(t))\exp(-\alpha_i\phi(i,t))\int_0^t e(i,s)\,ds\Big\}dt\Big].$$

From this equation, we obtain the kernel of the Gateaux differential as follows:

$$d(\boldsymbol{\phi},T,k) = -p_i\alpha_i\Big[R_{ik}(T)\exp(-\alpha_i\phi(i,T)) + \int_t^T (C+R_i f_i(z))\exp(-\alpha_i\phi(i,z))\,dz\Big].$$

Since $F(\boldsymbol{\phi}) = -f(\boldsymbol{\phi},T,k)$ and $c_0(i,t) = c_i$ in Lemma 2.2, we obtain Eqs. (3.11a) and (3.11b) as a necessary condition for the conditionally optimal $\boldsymbol{\phi}^*$.

To proof the sufficiency, the convexity of $f(\boldsymbol{\phi},T,k)$ with respect to $\boldsymbol{\phi}$ is proved as follows. We consider arbitrary effort allocations $\boldsymbol{\phi}^1$ and $\boldsymbol{\phi}^2$ that satisfy the constraints in Eq.(3.10) and set $\boldsymbol{\phi} = (1-\theta)\boldsymbol{\phi}^1 + \theta\boldsymbol{\phi}^2$, $0 \leqq \theta \leqq 1$. From Eq.(3.9), we obtain

$$\begin{aligned} f(\boldsymbol{\phi},T,k) = \Sigma_i p_i\Big[&R_{ik}(T)\exp\{-\alpha_i((1-\theta)\phi^1(i,T)+\theta\phi^2(i,T))\} \\ &- R_i(1-\beta_i)+\int_0^T (C+R_i f_i(t)) \\ &\times \exp\{-\alpha_i((1-\theta)\phi^1(i,t)+\theta\phi^2(i,t))\}dt\Big]. \end{aligned}$$

Since the negative exponential function is convex, we have the following relation for $R_{ik}(T) \geqq 0$.

$$\begin{aligned} f(\boldsymbol{\phi},T,k) \leqq \Sigma_i p_i\Big[&R_{ik}(T)\{(1-\theta)\exp(-\alpha_i\phi^1(i,T)) \\ &+ \theta\exp(-\alpha_i\phi^2(i,T))\} - R_i(1-\beta_i) \\ &+ \int_0^T (C+R_i f_i(t))\{(1-\theta)\exp(-\alpha_i\phi^1(i,t)) \\ &+ \theta\exp(-\alpha_i\phi^2(i,t))\}dt\Big] \\ = (1-\theta)&f(\boldsymbol{\phi}^1,T,k) + \theta f(\boldsymbol{\phi}^2,T,k). \end{aligned}$$

The equality holds only when $\theta = 0$ or $\theta = 1$, therefore $f(\boldsymbol{\phi}, T, k)$ is strictly convex with respect to $\boldsymbol{\phi}$.

Since the constraint is given by Eq.(3.10), $\boldsymbol{\phi}$ is an element of $\boldsymbol{\Psi}_0$ for all t. Therefore, from Lemma 2.2, the necessary conditions given by Eqs.(3.11a) and (3.11b) are also sufficient for $\boldsymbol{\phi}^*$ to be optimal. □

We obtain the following corollary by setting $t = T$ in Theorem 3.14.

Corollary 3.15. *If* $\phi^*(i, T) > 0$,

(3.12a) $$\frac{\alpha_i p_i}{c_i} R_{ik}(T)\exp(-\alpha_i \phi^*(i, T)) = \lambda(T),$$

if $\phi^*(i, T) = 0$,

(3.12b) $$\frac{\alpha_i p_i}{c_i} R_{ik}(T)\exp(-\alpha_i \phi^*(i, T)) \leqq \lambda(T).$$

$\lambda(t)$ in Theorem 3.14 has the following properties.

Corollary 3.16. *The function* $\lambda(t)$ *in Theorem* 3.14 *is a continuous and strictly decreasing function of* t, *and if* $f_i(t)$ *is a decreasing function of* t, $\lambda(t)$ *is a strictly convex function.*

Proof. Suppose $0 \leqq t_1 < t_2 \leqq T$. We can always find boxes, say boxes i and j, such that $\phi^*(i, t_1) > 0$ and $\phi^*(j, t_2) > 0$. Applying Theorem 3.14 at t_1 and t_2, we obtain

$$\frac{\alpha_j p_j}{c_j} \int_{t1}^{t2} (C + R_j f_j(z))\exp(-\alpha_j \phi^*(j, z))\,dz \leqq \lambda(t_1) - \lambda(t_2)$$
$$\leqq \frac{\alpha_i p_i}{c_i} \int_{t1}^{t2} (C + R_i f_i(z))\exp(-\alpha_i \phi^*(i, z))\,dz.$$

From this relation, we can conclude $\lambda(t_1) > \lambda(t_2)$ for $t_1 < t_2$, and in the limit as $(t_2 - t_1)$ approaches zero, $\lambda(t_1) = \lambda(t_2)$. Therefore, $\lambda(t)$ is a strictly decreasing and continuous function of t.

The convexity of $\lambda(t)$ is proved as follows. We consider t such that $t = (1-\theta)t_1 + \theta t_2$, $0 \leqq \theta \leqq 1$, $0 \leqq t_1 < t_2 \leqq T$. Then there always exists a box, box i, such that $\phi^*(i, t) > 0$. From Theorem 3.14, we obtain

$$\begin{aligned}
\lambda(t) &= \frac{\alpha_i p_i}{c_i}[R_{ik}(T)\exp(-\alpha_i \phi^*(i, T)) \\
&\qquad + \int_t^T (C + R_i f_i(z))\exp(-\alpha_i \phi^*(i, z))\,dz] \\
&= (1-\theta)\frac{\alpha_i p_i}{c_i}[R_{ik}(T)\exp(-\alpha_i \phi^*(i, T)) \\
&\qquad + \int_{t1}^{T} (C + R_i f_i(z))\exp(-\alpha_i \phi^*(i, z))\,dz] \\
&\qquad + \theta \frac{\alpha_i p_i}{c_i}[R_{ik}(T)\exp(-\alpha_i \phi^*(i, T)) \\
&\qquad + \int_{t2}^{T} (C + R_i f_i(z))\exp(-\alpha_i \phi^*(i, z))\,dz] \\
&\qquad + \frac{\alpha_i p_i}{c_i}[\theta \int_t^{t2} (C + R_i f_i(z))\exp(-\alpha_i \phi^*(i, z))\,dz \\
&\qquad - (1-\theta)\int_t^{t1} (C + R_i f_i(z))\exp(-\alpha_i \phi^*(i, z))\,dz].
\end{aligned}$$

Therefore, we have

$$\lambda(t) \leqq (1-\theta)\lambda(t_1) + \theta\lambda(t_2) + \frac{\alpha_i p_i}{c_i}\Big[\theta\int_t^{t2}(C+R_if_i(z))\exp(-\alpha_i\varphi^*(i,z))dz - (1-\theta)\int_t^{t1}(C+R_if_i(z))\exp(-\alpha_i\varphi^*(i,z))dz\Big].$$

The third term in the right-hand side is not positive if $f_i(t)$ is a decreasing function of t. Therefore, $\lambda(t) \leqq (1-\theta)\lambda(t_1)+\theta\lambda(t_2)$, where the equality holds only for $\theta=0$ or $\theta=1$. Thus $\lambda(t)$ is a strictly convex function of t if $f_i(t)$ is a decreasing function of t. □

Corollary 3.17. *If both $\phi^*(i,t)$ and $\phi^*(j,t)$ are positive in $t \in [t_1, t_2]$, the following relation holds at any t in $[t_1, t_2]$.*

$$\frac{\alpha_i p_i}{c_i}\{C+R_if_i(t)\}\exp(-\alpha_i\varphi^*(i,t)) = \frac{\alpha_j p_j}{c_j}\{C+R_jf_j(t)\}\exp(-\alpha_j\varphi^*(j,t)).$$

Proof. Applying Theorem 3.14 to both t and $t+\Delta t$, $t_1\leqq t$, $t+\Delta t\leqq t_2$, we have

$$\lambda(t)-\lambda(t+\Delta t) = \frac{\alpha_i p_i}{c_i}\{C+R_if_i(t)\}\exp(-\alpha_i\varphi^*(i,t))\Delta t = \frac{\alpha_j p_j}{c_j}\{C+R_jf_j(t)\}\exp(-\alpha_j\varphi^*(j,t))\Delta t.$$

Therefore, Corollary 3.17 is proved. □

Remark 3.5. We examine the physical meaning of the condition of Theorem 3.14. Let $\zeta(i,t,s)$ be the searching effort allocated to box i in $[t,s]$; $\zeta(i,t,s) = \int_t^s \phi(i,z)dz$. Then, Eqs.(3.11a) and (3.11b) in Theorem 3.14 are rewritten as follows: if $\phi^*(i,t) >(=) 0$,

$$\frac{\alpha_i p_i\exp(-\alpha_i\varphi^*(i,t))}{c_i}\Big[R_i(1-F_i(t))+\{R_{ik}(T)\exp(-\alpha_i\zeta^*(i,t,T)) - R_i(1-F_i(t)) + \int_t^T(C+R_if_i(z))\exp(-\alpha_i\zeta^*(i,t,z))dz\}\Big] =(\leqq) \lambda(t).$$

The numerator of the first term, $\alpha_i p_i\exp(-\alpha_i\varphi^*(i,t))$, is the detection probability rate for the target by the allocation of unit searching effort density to box i at t. The first term in the brackets, $R_i(1-F_i(t))$, is the expected reward gained by detection of the target at t in box i. The second term,

$$\{R_{ik}(T)\exp(-\alpha_i\zeta^*(i,t,T))-R_i(1-F_i(t)) + \int_t^T(C+R_if_i(z))\exp(-\alpha_i\zeta^*(i,t,z))dz\},$$

is the conditional risk of the whereabouts search in $[t,T]$ given that the target is in box i, as shown by the term in the brackets in Eq.(3.9).

Since the search is to be continued until T_k if the target is not detected and box k is to be guessed, the second term is interpreted as the saved risk earned by detecting the target in box i at time t. Therefore, the terms in the brackets are interpreted as the return that motivates the search in box i at t. Thus the terms in the brackets multiplied by the numerator of the first term mean the expected marginal return in box i at t. Hence, the left-hand side of Eqs.(3.11a) and (3.11b) is the ratio of the expected marginal return to the cost with the unit density of searching effort allocated to box i at t. Therefore, the physical meaning of Theorem 3.14 is stated as follows. If box i is to be searched at t, the amount of the searching effort should be determined in such a way that the expected marginal return versus cost ratio is balanced to $\lambda(t)$ among the boxes being searched at t. If box i is not to be searched at t, the reason is that box i does not have a larger expected marginal return versus cost ratio than $\lambda(t)$.

Next, we will consider the conditionally optimal stopping time T_k^* given that $\boldsymbol{\phi}_T^*$ obtained from Theorem 3.14 is employed up to T and box k is guessed when the search $\boldsymbol{\phi}_T^*$ fails to detect the target. The stopping time that minimizes $f(\boldsymbol{\phi}_T^*, T, k)$ with respect to T is sought.

<u>Theorem 3.18.</u> *A necessary condition for the conditionally optimal stopping time T_k^* is*

$$(3.13)\qquad Q(T_k^*) + \frac{p_k R_k^0 f_k(T_k^*)}{C}\exp(-\alpha_k \varphi_T^*(k, T_k^*)) = \lambda(T_k^*),$$

where $Q(T_K^*) = \Sigma_i p_i \exp(-\alpha_i \varphi_T^*(i, T_k^*))$ is the nondetection probability.

<u>Proof.</u> Since the derivative of $\varphi(i, t)$ with respect to t at $t = T$ is defined by the left derivative and $F_i(t)$ is assumed to be differentiable, $f(\boldsymbol{\phi}, T, k)$ is also differentiable with respect to T. Using Eq. (3.12a), we obtain

$$\frac{\partial f(\boldsymbol{\phi}, T, k)}{\partial T} = p_k R_k^0 f_k(T)\exp(-\alpha_k \varphi(k, T)) - \lambda(T)C + CQ(T).$$

Eq.(3.13) is derived from $\partial G(\boldsymbol{\phi}, T, k)/\partial T = 0$ at $T = T_k^*$. □

The following theorem states the conditions for the search to be continued or not.

<u>Theorem 3.19.</u> 1. *The search should not be stopped at T with guessing box k, if*

$$(3.14)\qquad \max_i \frac{\alpha_i p_i(T) R_{ik}(T)}{c_i} - \frac{p_k(T) R_k^0 f_k(T)}{C} > 1.$$

2. *The search should not be continued until T, if*

$$(3.15)\qquad \max_i \frac{\alpha_i p_i(T) R_{ik}(T)}{c_i} - \frac{p_k(T) R_k^0 f_k(T)}{C} < 1,$$

where $p_i(T)$ is the posterior probability of the target being in box i at T given that the conditionally optimal search plan $\boldsymbol{\phi}_T^*$ fails to detect the target.

$$p_i(T) = \frac{p_i \exp(-\alpha_i \phi_T^*(i, T))}{Q(T)}.$$

Proof. Let $\boldsymbol{\phi}_T^*$ be the conditionally optimal search plan given T and k. We consider an extended search plan $\boldsymbol{\phi}_{T+\varepsilon}$ such that $\phi_{T+\varepsilon}(i, t) = \phi_T^*(i, t)$, $t \in [0, T]$ and $\phi_{T+\varepsilon}(i, t) = \phi_T^*(i, T)$, $t \in (T, T+\varepsilon]$, where ε is assumed to be a positive and sufficiently small time interval. Then we have

$$\begin{aligned} f(\boldsymbol{\phi}_{T+\varepsilon}^*, T+\varepsilon, k) &- f(\boldsymbol{\phi}_T^*, T, k) \\ &\leq f(\boldsymbol{\phi}_{T+\varepsilon}, T+\varepsilon, k) - f(\boldsymbol{\phi}_T^*, T, k) \\ &= -\varepsilon CQ(T)[\max_i \frac{\alpha_i p_i(T) R_{ik}(T)}{c_i} - \frac{p_k(T) R_k^0 f_k(T)}{C} - 1] + o(\varepsilon). \end{aligned}$$

Therefore, if the inequality in Eq. (3.14) holds, $f(\boldsymbol{\phi}_{T+\varepsilon}^*, T+\varepsilon, k) < f(\boldsymbol{\phi}_T^*, T, k)$, and the search should be extended at least to $T+\varepsilon$.

Next, we will truncate $\boldsymbol{\phi}_T^*$ at $T-\varepsilon$. Then we have

$$\begin{aligned} f(\boldsymbol{\phi}_{T-\varepsilon}^*, T-\varepsilon, k) &- f(\boldsymbol{\phi}_T^*, T, k) \\ &\leq f(\boldsymbol{\phi}_T^*, T-\varepsilon, k) - f(\boldsymbol{\phi}_T^*, T, k) \\ &= \varepsilon CQ(T)[\max_i \frac{\alpha_i p_i(T) R_{ik}(T)}{c_i} - \frac{p_k(T) R_k^0 f_k(T)}{C} - 1] + o(\varepsilon). \end{aligned}$$

Hence, under the condition in Eq. (3.15), $f(\boldsymbol{\phi}_{T-\varepsilon}^*, T-\varepsilon, k) < f(\boldsymbol{\phi}_T^*, T, k)$. Then the search should be stopped before T. □

Since $\lambda(T)$ is given by Eqs. (3.12a) and (3.12b), we can easily derive the following corollary from Theorems 3.18 and 3.19.

Corollary 3.20. *We define a function G(T) by*

$$G(T) = \max_i \frac{\alpha_i p_i(T) R_{ik}(T)}{c_i} - \frac{p_k(T) R_k^0 f_k(T)}{C}. \tag{3.16}$$

A necessary condition for the conditionally optimal stopping time T_k^* is

$$\begin{aligned} G(T) &\geq 1 \quad \text{if } T < T_k^*, \\ G(T) &= 1 \quad \text{if } T = T_k^*, \\ G(T) &\leq 1 \quad \text{if } T > T_k^*, \end{aligned}$$

for any T in the neighborhood of T_k^*.

Corollary 3.20 corresponds to Corollary 3.7 on the optimal stopping time of search for an immortal target. As discussed later, the second term of the right-hand side in Eq. (3.16) reflects the mortality of the target.

The following corollary is easily derived from Corollary 3.20.

Corollary 3.21. *A necessary condition for* $T_k^* = 0$ *is*

$$\max_i \frac{\alpha_i p_i}{c_i} \{W_k + (1-\beta_i) R_i - \delta_{ik}(1-\beta_k) R_k^0\} - \frac{p_k R_k^0 f_k(0)}{C} < 1. \quad \square$$

Corollary 3.22. *We assume that there exists a time* t_0 *such that* $f_i(t) = 0$ *for all* i *and* $t \geqq t_0$. *Then a necessary condition for* $T_k^* = \infty$ *is*

$$\max_i \gamma_i (R_i - \delta_{ik} R_k^0) > \Sigma_i \frac{c_i}{\alpha_i} - W_k. \tag{3.17}$$

If the inequality sign in Eq. (3.17) *is reversed,* T_k^* *is finite.*

Proof. Suppose $T_k^* = \infty$. The search after t_0 is identical with the search for an immortal stationary target. Then, as shown by Corollary 1.6, there exists a time point T^0,

$$T^0 = t_0 + \frac{1}{C} \Sigma_i \frac{c_i}{\alpha_i} \{\log \frac{\alpha_i p_i(t_0)/c_i}{\alpha_j p_j(t_0)/c_j}\},$$

and for any $t \geqq T^0$, the conditionally optimal search plan $\phi^*(i, t)$ and the posterior probability of the target $p_i(t)$ become

$$\phi^*(i, t) = \frac{C}{\alpha_i \Sigma_m (c_m/\alpha_m)}, \tag{3.18a}$$

$$p_i(t) = \frac{c_i/\alpha_i}{\Sigma_m (c_m/\alpha_m)}, \tag{3.18b}$$

for all i and $t \geqq T^0$, where j is the box having the minimum value of $\alpha_i p_i(t_0)/c_i$. Substituting $f_i(\infty) = 0$, $R_{ik}(\infty) = W_k + \gamma_i (R_i - \delta_{ik} R_k^0)$, and Eqs. (3.18a) and (3.18b) to Eq. (3.14), we obtain Eq. (3.17). If the inequality sign in Eq. (3.17) is reversed, Eq. (3.15) holds at $T = T^0$, and therefore, $T_k^* < T^0$ is concluded. $\square$

Remark 3.6. Theorem 3.18 and Corollary 3.20 explain the behavior of the expected marginal return versus cost ratio at the neighborhood of the optimal stopping time T_k^*. By the definition, $R_{ik}(T)$ is the conditional value of the expected return at T in box i given that the target is contained in box i and box k is guessed. Hence, the first term of the right-hand side of Eq. (3.16) means the expected marginal return versus cost ratio by unit density of searching effort in box i at T. Meanwhile, the numerator of the second term is the expected lost reward by the death of the target at T if the search is continued and guessing box k is delayed by unit time and the denominator is the searching cost used in unit time. Therefore, the second term means the expected lost price versus cost ratio at T if the stopping time of the search is prolonged by unit time. Hence, the right-hand side of Eq. (3.16) is interpreted as the ratio of the marginal expected return to cost if the search is continued by unit time at T. Therefore, the interpretation of Corollary 3.20 is as follows: The expected marginal return versus cost ratio of the search being continued decreases across unity as the time to guess increases across the conditionally optimal stopping time T_k^*. Box k should be guessed at the instant when the expected marginal return becomes equal to the cost.

The final stage of our optimization of the whereabouts search for a mortal target is determining the optimal box k^* to be guessed, such that $f(\phi_{Tk}^*, T_k^*, k)$ is minimized. Then the triplet $(\phi_{Tk*}^*, T_{k*}^*, k^*)$ is the optimal whereabouts search plan.

Theorem 3.23. *A necessary condition for the optimal box k^* to be guessed is*

(3.19) $$W_{k*}-p_{k*}(T_{k*}^*)R_{k*}^0(1-F_{k*}(T_{k*}^*)) \leqq W_j-p_j(T_{k*}^*)R_j^0(1-F_j(T_{k*}^*))$$
for all j.

Proof. Suppose $(\phi_{Tk*}^*, T_{k*}^*, k^*)$ is optimal and consider a whereabouts search plan $(\phi_{Tk*}^*, T_{k*}^*, j)$, $j \neq k^*$. Then we have

$$\begin{aligned} f(\phi_{Tk*}^*, T_{k*}^*, k^*) &- f(\phi_{Tk*}^*, T_{k*}^*, j) \\ &= Q(T_{k*}^*)[\ W_{k*}- p_{k*}(T_{k*}^*)R_{k*}^0(1-F_{k*}(T_k^*)) \\ &\qquad - W_j + p_j(T_{k*}^*)R_j^0(1-F_j(T_{k*}^*))]. \end{aligned}$$

Since $(\phi_{Tk*}^*, T_{k*}^*, k^*)$ is optimal, the left-hand side of this equation is not positive, and therefore the theorem is established. □

Remark 3.7. Theorem 3.23 states a property of the optimal box k^* to be guessed. Since the term $W_i-p_i(T)R_i^0(1-F_i(T))$ is the expected risk of guessing box i at T, the inequality in Eq.(3.19) means that if box k^* is the optimal box to be guessed, the smallest risk is obtained by guessing box k^* at T_k^*. Thus Theorem 3.23 elucidates the property of the optimal box to be guessed at T_{k*}^* clearly, but this theorem is not so powerful as to determine the optimal box k^* directly. We may have to evaluate $f(\phi_{Tk}^*, T_k^*, k)$ for all boxes and select the minimum value among them to determine the optimal box to be guessed.

Remark 3.8. In this section, we deal with the whereabouts search model for a mortal target under the following assumptions: (1) continuous time space, (2) discrete target space, (3) continuous divisibility of the searching effort, and (4) exponential detection function. We shall consider modifications of these assumptions.

The continuous time space is easily modified to a discrete time space. In the discrete-time version, most of the results derived here are valid if the integration with respect to t is replaced by the summation. However, the equality signs in Eq.(3.13) do not generally hold by the discontinuity of time.

The assumption of the discrete target space is essential for the definition of guessing. If the target space is continuous, we must introduce another definition of guessing.

The continuous divisibility of the searching effort is a very important assumption for our model in simplifying the problem. If the searching effort is quantized, the condition for the optimal search plan will be drastically complicated.

Finally, the assumption of the exponential function for the conditional detection probability is not essential to deal with the problem. The model is easily generalized by interchanging the exponential function with a regular detection function $b(i, \varphi)$ which is a strictly increasing, concave, differentiable function of φ. By replacing $\exp(-\alpha_i\varphi)$ and $\alpha_i\phi\exp(-\alpha_i\varphi)$ with $1-b(i, \varphi)$ and $\partial b(i, \varphi)/\partial t$, respectively, the theorems derived in this section are valid.

Remark 3.9. The model dealt with here is considered a generalized model of the previous studies by several authors. (1) If we set $F_i(t)=0$ for all i and t, the model dealt with here is completely identical to the model investigated by Iida and Kanbashi (1983), which is the whereabouts search model for an immortal target described in the previous section. (2) If $W_k=0$ and $R_k{}^0=0$ for all k, the concept of guessing loses its actual meaning. Then the whereabouts search model dealt with here is identical with the detection search model for a target with a random lifetime, which was investigated by Nakai (1982). (3) If we set $F_i(t)=0$ together with $W_i=R_i{}^0=0$ for all i, the situation becomes the detection search for a stationary immortal target. This model was studied by Iida (1984), and is described in Chapter 1.

If the parameters are chosen as stated in our model, all results presented in this section completely coincide with the results of these papers. For example, if we set $F_i(t)=0$ for all i and t (in other words, the target is immortal), the conditions of Theorem 3.14 become if $\phi^*(i, t)$ >(=) 0,

(3.20)
$$\frac{\alpha_i p_i}{c_i}\Big[R_{ik}\exp(-\alpha_i\varphi^*(i, T)) + \int_t^T C\exp(-\alpha_i\varphi^*(i, z))dz\Big] =(\leqq)\ \lambda(t).$$
$$R_{ik} = R_i + W_k - \delta_{ik}R_k{}^0.$$

These conditions are identical to those of Theorem 3.1 derived in the previous section. Iida and Kanbashi (1983) gave an algorithm for calculating the conditionally optimal search plan using Eq.(3.20), which is a modified procedure of the FAB algorithm proposed by Washburn (1983). This algorithm is also applicable to our whereabouts search model for a mortal target by replacing R_{ik} and C in Eq.(3.20) with $R_{ik}(T)$ and $C+R_if_i(t)$, respectively; therefore, we can easily calculate the conditionally optimal search plan numerically.

3.4. Numerical Examples

In this section, we will examine several simple examples of the whereabouts search in two boxes to see the influence of the lifetime of the target on the optimal plan. Case 1 is an example of the whereabouts search for an appeared (immortal) target, which was discussed in Section 3.2 and has parameter values shown in Table 3.1.

Table 3.1. The parameters of Case 1.

i	p_i	α_i	R_i	R_i^0	c_i	C	W_i	$F_i(t)$
1	0.5	1	5	10	1	1	1	0
2	0.5	1	10	5	1		1	

The conditionally optimal expected risks, $f(\phi^*, T, k)$, k=1,2, given the stopping time T and box k to be guessed are shown in Fig. 3.1. The abscissa of this figure is the stopping time T.

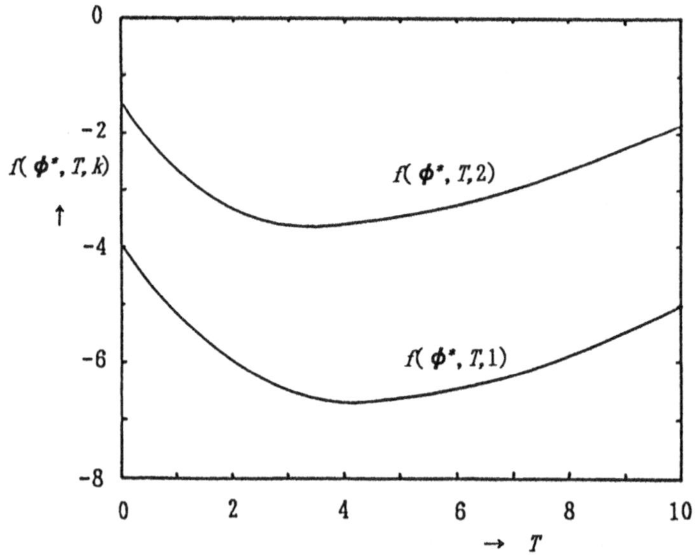

Figure. 3.1. $f(\phi^*, T, k)$.

In Fig. 3.1, the minimum value, -6.73, occurs on the curve $f(\phi^*, T, 1)$ at T = 3.88. Since the point of the minimum value of this figure corresponds to the optimal whereabouts search plan, the optimal box to be guessed is k^* = 1 and the optimal stopping time is T^* = 3.88. The optimal distribution of searching effort is calculated as

$$\phi^*(1, t) = 0, \quad \phi^*(2, t) = 1, \qquad \text{for} \quad 0 \leqq t \leqq 3.88,$$

namely, the optimal whereabouts search is: only box 2 is searched until $t = 3.88$ and box 1 is guessed at the end of the search if the target is not detected. In this case, as shown in Table 3.1, the reward when the target is detected in box 2 is twice as much as box 1, and therefore, box 2 is worth searching. On the other hand, since the reward given by the correct guess in box 1 is twice as much as box 2, box 1 is appropriate to be guessed. Hence, box 2 should be searched and box 1 should be guessed; the optimal whereabouts search plan is natural intuitively.

The next case, Case 2, is an example of the whereabouts search for the mortal target investigated in Section 3.3. We assume the same parameters as Case 1, except for the lifetime of the target in box 1. The distribution of lifetime of the target in box 1 is assumed to be an exponential distribution function:

$$F_1(t) = 1 - \exp(-\mu_1 t), \quad \mu_1 = 0.2 \quad \text{and} \quad \beta_1 = \gamma_1 = 0,$$

and the target in box 2 is assumed to be immortal, namely,

$$\mu_2 = \beta_2 = \gamma_2 = 0.$$

The conditional expected risk is calculated and shown in Fig. 3.2. From this figure, we have the optimal whereabouts search plan as

$$k^* = 1, \qquad T^* = 3.50, \qquad f(\boldsymbol{\phi}^*, 3.50, 1) = -5.94.$$

The optimal distribution of the searching effort $\boldsymbol{\phi}^*$ is shown in Fig. 3.3.

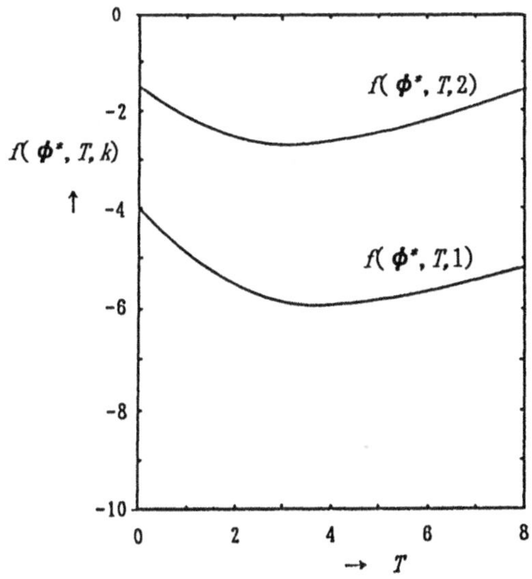

Figure 3.2. $f(\boldsymbol{\phi}^*, T, k)$.

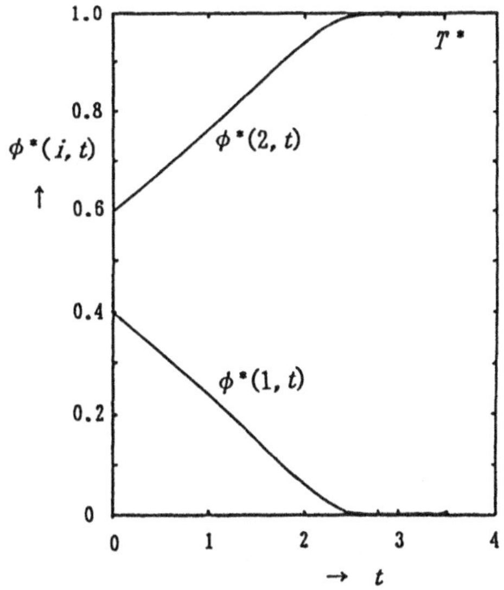

Figure 3.3. $\boldsymbol{\phi}^*$.

The distinctive feature of the optimal plan of Case 2 is the distribution of searching effort, $\boldsymbol{\phi}_{T_{k*}}^*$: box 1 is searched from the start of the search, but it is never searched in Case 1. Moreover, the expected risk increases and the search time is shortened compared to Case 1. These properties are explained as follows. Since the target in box 1 is mortal in Case 2, box 1 must be searched early to detect the target alive, and since the possibility of target death increases the risk of the search, the search for the mortal target should not be continued for long.

The next case, Case 3, examines the optimal whereabouts search plan for the target with a shorter lifetime than that of Case 2. In Case 3, the death rate μ_1 of the target in box 1 is changed from 0.2 to 1.2; the other parameters remain the same as Case 1. In this case, the optimal whereabouts search plan is

$$k^* = 1, \quad T^* = 1.18, \quad f(\boldsymbol{\phi}^*, 1.18, 1) = -4.52,$$

and the optimal distribution of the searching effort is shown in Fig. 3.4. By comparing the optimal plan of Case 3 with those of Case 1 and 2, we can see that the searching effort is more concentrated to box 1 at the early stage of the search and the search is stopped earlier. Varying the death rate μ_1 of the target from 0 to 1.2, we evaluate the optimal stopping time T_{k*}^* and the optimal searching effort at the beginning of the search $\phi^*(1,0)$ and show the results in Fig. 3.5.

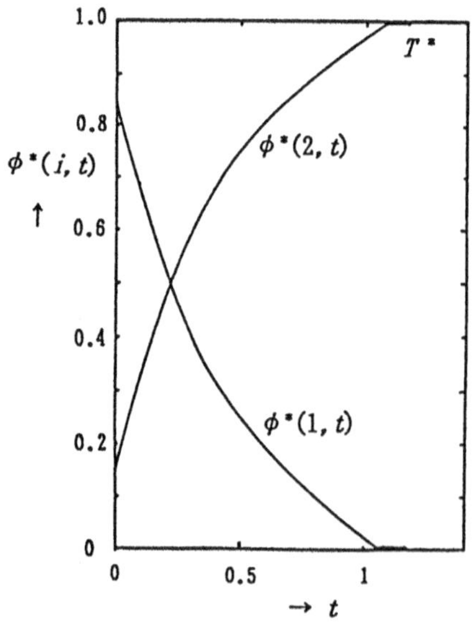

Figure 3.4. $\boldsymbol{\phi}^*$.

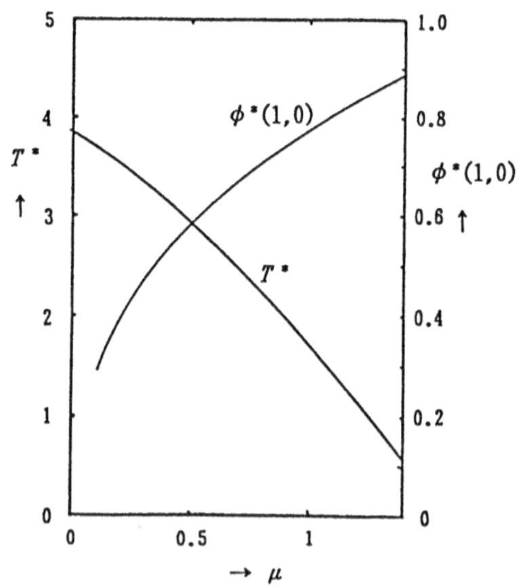

Figure 3.5. T^* and $\phi^*(1,0)$.

From the figure, we can conclude that the more the death rate increases in box 1, the more the searching effort at the first stage of the search is concentrated to box 1 and the optimal stopping time is shortened.

To see the influence of the death rate on the optimal whereabouts search plan clearly, in Case 4, we set both the rewards of the detection of the target and the correct guess equal for all boxes; $R_i = R_i^0 = 5$, whereas the other parameters are kept the same as Case 2. The parameters are shown in Table 3.2.

Table 3.2. The parameters of Case 4.

i	p_i	α_i	R_i	R_i^0	c_i	C	W_i	μ_i
1	0.5	1	5	5	1	1	1	0.2
2	0.5	1	5	5	1		1	0

In this case, the boxes are equally characterized except the target is mortal in box 1 and immortal in box 2. Therefore, the optimal search plan is directly influenced by the type of target. The optimal whereabouts search policy for Case 4 is:

$$k^* = 2, \quad T^* = 2.29, \quad f(\phi^*, 2.29, 2) = -2.17,$$
$$\phi^*(1, t) = 1 \text{ and } \phi^*(2, t) = 0 \quad \text{for all } t;\ 0 \leqq t \leqq 2.29.$$

In this case, the optimal guess box is changed from box 1 to box 2. All the searching effort is allocated to box 1 to detect the mortal target in the early stage of the search and at the end of the search, box 2 is guessed to detect the immortal target.

CHAPTER 4

Optimal Investigating Search Plan for Contacts in Two-Stage Search

4.1. Introduction

In the previous sections, we have investigated the optimal search plan in various search situations without considering the possibility of false contacts. In this chapter, a noisy search model will be dealt with. In the noisy search, the detection device of the searcher sometimes gives false alarms of the target, and hence the second stage of search is needed to ascertain whether the detection is true or not. The first- and the second-stage search are called the broad search and the investigating search, respectively.

The optimal search plan in a two-stage search was first studied by Engel (1957). In his model, contacts obtained in the broad search need not be immediately investigated. After the broad search, a set of contacts is selected by an "(n, s) clustering technique" from the contacts gained, and then only the selected contacts are investigated in the investigating search. The (n, s) clustering technique is specified by the number of covering of the broad search (n) and the minimum number of contacts within a specified range from the center of the cluster (s). An optimal balance of searching effort between the first- and the second-stage search were obtained. Smith and Walsh (1971) dealt with a noisy search model where both the target space and the time space are assumed discrete. Attention is paid to the first stage of the search in their model. Stone and Stanshine (1971) investigated an optimal broad search plan for the two-stage search minimizing the expected time to find the target assuming the contact investigation must not be interrupted until the contact is identified. This study was generalized by Stone et al. (1972) and the assumption of the uninterrupted contact investigation is removed. Assuming the presence of Poisson-distributed false targets, they derived an optimal search plan that minimizes the expected time to find the target. They obtained the optimal search plan that specifies both the distributions of the broad search effort and the contact investigation effort as functions of the location and the broad search time. Dobbie (1973) studied a search problem in the presence of false contacts and showed that the optimal solution is contingent on the number of false targets found and the locations and times of contact.

The distinctive feature of these studies except the one by Smith and Walsh is that they deal with false contacts originated from signals emitted by real objects that are very similar to those from the true target. The false contacts of this type can be distinguished in the second stage of search and can be marked false when located.

However, if the contact is of the false alarm type originated from background noise, there is no object to be located, usually no positive information telling it to be a false contact is obtained through investigation. This type of false contact problem was first studied by Kisi (1979). He dealt with noise-type false contacts and considered the optimal stopping time of the investigating search depending on the probability of the contact being true. A necessary condition for the optimal stopping time of the investigating search that minimizes the expected time of detecting the target is derived and the case of exponential detection law in the investigating search is examined in detail. Iida (1989a) generalized Kisi's model with respect to the characteristics of the contact. The search model dealt with in this chapter is based on Iida and the search situation is as follows.

A searcher is searching for targets in a given area. The search operation begins with the broad search, where the searcher will get information telling the position of a possible target sooner or later. A piece of information of this kind is called a contact and the target position reported in the contact is a datum point.

A contact may be or may not be true. Coming signals are sometimes just like the signals from the target and difficult to distinguish. The searcher is unable to know whether or not the contact is true, but knows the probability of the contact being true. There is also another uncertainty. The datum point reported is usually erroneous even if the contact is true, since the search detector cannot be free from measurement error. The datum point error has influence on the possibility of detecting the target in the investigating search. If the error distance between the true target position and the datum point is large, the investigating search might fail to locate the target. Finally a third factor should be taken into account. When a contact is obtained, usually it takes time to estimate the probability of the contact being true and to evaluate the parameters of the error distribution of the datum point. If the time lag needed to evaluate the contact is short, the investigating search is started soon, and successful location of the target is expected. On the contrary, if considerable time is needed before the start of the investigating search, we cannot expect good news since the target has much time for evasive motion to upset the search plan for investigation. Thus, we see a spectrum of quality of contacts from high to low. If the probability of the contact being true is high and the datum point error is small with a short time lag, the contact is called of high quality. A contact with a low probability of telling truth, a large datum point error, and a long time lag is called of low quality.

Upon arriving at a contact, the searcher evaluates its quality and decides whether or not to investigate it. If the quality is low, it should be disregarded. If it is sufficiently high, the contact should be investigated. Then the searcher switches his effort to the investigating search, and the search will be in the second stage. If the investigating search is continued with no detections, the contact under investigation get to be suspicious and the hope for successful location of the target becomes dimmer. The searcher should stop the investigating search at some appropriate time and return to the broad search to get another contact. When to stop the investigating search is hence an important problem to be solved.

In the next section, the problem is formulated as a fractional programming problem and necessary and sufficient conditions for the optimal selection of contacts for investigation and the optimal stopping time of the investigating search are derived. In a later section, several numerical examples will be given and properties of the optimal investigating search are discussed.

4.2. Optimal Investigating Search Plan

4.2.1. Optimal Selection of Contacts for Investigation and Optimal Stopping Time

The model of the two-stage search process dealt with here is described in detail as follows.

1. A searcher is searching for targets in a given area. The search operation begins with the broad search from $t = 0$. When the searcher gains a contact in the broad search, he evaluates the quality of the contact. If it is worth investigating, he switches his effort immediately to the investigating search to confirm whether the contact is true or not.

2. The arrival of the contacts in the broad search is assumed to be a Poisson process with rate α that does not change in time and is not influenced by past contacts.

3. A contact is specified by an information triplet (p, s, d).

(a) p, $0 < p \leqq 1$, is the probability of the contact being true. (b) s, $0 < s < \infty$, is the time needed to evaluate the quality of the contact. It is the time lag from the contact to the beginning of the investigating search. We assume that the searcher must not stop the evaluation of the contact before knowing the value of (p, d). The time lag s is wasted even when the searcher disregards the contact. (c) d, $0 \leqq d < \infty$, is an appropriate parameter that stands for the accuracy of the reported datum point (called the datum point error). The contact with a large value of d has low accuracy. We assume that the searcher gets the correct value of the triplet (p, s, d) when a contact is evaluated.

4. The a priori probability density of a contact being (p, s, d) is given by $dB(p, s, d)$ and is known to the searcher.

5. The conditional probability of detecting the target in the investigating search until t given that the contact (p, s, d) is true is given by a function $H_{sd}(t)$, where the net time for investigating the contact is $(t-s)$ since s is wasted before the start of the investigating search. This statement implicitly assumes that the standard way of the investigating search around the datum point is established and assumes that the conditional detection probability can be estimated taking the time lag, the accuracy of the datum point, and the method of investigating search into account. We assume the following properties for the conditional detection function $H_{sd}(t)$. (a) $H_{sd}(t) = 0$, $0 \leqq t \leqq s$. (b) $H_{sd}(t)$ is a continuous, strictly increasing, and concave function of t for $t > s$. (c) $H_{sd}(t)$ is differentiable twice for all t $(>s)$. The first- and the second-order derivatives of $H_{sd}(t)$ are denoted by $h_{sd}(t)$ and $h_{sd}'(t)$ respectively. We define the derivative of $H_{sd}(t)$ at s by

$$h_{sd}(s) = \lim_{t \to s+0} h_{sd}(t).$$

Therefore, by assumption (b), for $t \geqq s$, we have

$$h_{sd}(t) > 0, \quad h_{sd}(\infty) = 0, \quad \text{and} \quad h_{sd}'(t) < 0.$$

(d) $h_{sd}(s)$ is assumed to be nonincreasing with respect to both s and d.

6. If the contact is false, no positive information telling it to be a false contact is obtained by the investigating search. Hence, to avoid wasting time, the searcher should stop the investigating search when an unsuccessful search is continued too long. The stopping time of the investigating search for the contact (p, s, d) is denoted by T_{psd}. Since the time lag s is assumed to be wasted even when the contact (p, s, d) is disregarded, $T_{psd} \geqq s$. We denote $\mathbf{T} = \{T_{psd}\}$ and $\mathbf{T}$ is called an investigating search plan.

7. During the investigating search, the broad search activity is off and no new contacts are coming in. When the investigating search is stopped, the broad search is resumed without any time loss.

8. The measure of effectiveness for the investigating search plan

is defined as the expected time $E(\mathbf{T})$ from the beginning of the broad search to the first acquisition of the true target in the investigating search. The investigating search plan that minimizes the expected time $E(\mathbf{T})$ is called optimum and is denoted by $\mathbf{T}^*$.

Under these assumptions, we formulate the expected time for detecting the target, $E(\mathbf{T})$. From the start of the broad search, $1/\alpha$ will be elapsed in the mean before the first contact is gained, and the probability density of the first contact being (p, s, d) is given by $dB(p, s, d)$. Let the conditional expected time for detecting the target from the instant of a contact (p, s, d) be $E_{psd}(\mathbf{T})$. Then we have the following relation:

$$E(\mathbf{T}) = \frac{1}{\alpha} + \int E_{psd}(\mathbf{T})\, dB(p, s, d). \tag{4.1}$$

If no limits of integration are specified, it means integration over all the domain.

Now suppose a contact (p, s, d) is in hand and the investigating search is begun with stopping time T_{psd}. The target will be detected if the contact is true and if the investigating search ends in success in detecting the target before the stopping time T_{psd}. The probability of success is given by $pH_{sd}(T_{psd})$ and the conditional expected detection time is given by

$$\int_s^{T_{psd}} \frac{zh_{sd}(z)}{H_{sd}(T_{psd})}\, dz.$$

If the investigating search fails to detect the target until the stopping time T_{psd}, the broad search is resumed again; the expected time of detection from this time on is given by $E(\mathbf{T})$. Therefore, we have

$$E_{psd}(\mathbf{T}) = p\int_s^{T_{psd}} zh_{sd}(z)\, dz + (1-pH_{sd}(T_{psd}))(T_{psd}+E(\mathbf{T})). \tag{4.2}$$

Substituting Eq. (4.2) into Eq. (4.1), we obtain

$$E(\mathbf{T}) = \frac{N}{\alpha \int pH_{sd}(T_{psd})\, dB(p, s, d)}, \tag{4.3}$$

$$N = 1+\alpha \int \left[p\int_s^{T_{psd}} zh_{sd}(z)\, dz+(1-pH_{sd}(T_{psd}))\,T_{psd}\right] dB(p, s, d),$$

where $\mathbf{T} \neq \{T_{psd}=s \text{ for all } (p, s, d)\}$. Therefore, our problem is formulated as a fractional programming problem to find an optimal $\mathbf{T}^*$ that minimizes $E(\mathbf{T})$ given by Eq. (4.3) subject to restrictions

$$T_{psd} \geq s, \qquad \mathbf{T} \neq \{T_{psd} = s\}. \tag{4.4}$$

The set of $\{T_{psd}\}$ satisfying Eq. (4.4) is called a feasible search plan.

From the definition of $E(\mathbf{T})$ given by Eq. (4.3), the function $E(\mathbf{T})$ has the following properties.

Theorem 4.1. 1. *$E(\mathbf{T})$ becomes infinite when $\{T_{psd}\}$ approaches $\{s\}$.*

(4.5) $$\lim_{all\ T_{psd}\to s} E(\mathbf{T}) = \infty.$$

2. *If $\lim_{t\to\infty} pH_{sd}(t) < 1$, then*

(4.6) $$\lim_{T_{psd}\to\infty} E(\mathbf{T}) = \infty.$$

3. *$E(\mathbf{T})$ is continuous and differentiable to any direction at any inner point $\mathbf{T}^I$, $\mathbf{T}^I = \{T_{psd} | T_{psd} > s\}$.*

4. *$E(\mathbf{T})$ has a positive minimum value $E(\mathbf{T}^*)$.*

Proof. 1. If T_{psd} approaches s for all (p, s, d), the denominator and numerator of Eq.(4.3) approach zero and $\alpha \int s dB(p, s, d)+1$, respectively, since $\lim_{t\to s+0} H_{sd}(t) = 0$ by the definition. Therefore, Eq.(4.5) is derived. This means it is not optimal to disregard all the contacts.

2. If $\lim_{t\to\infty} pH_{sd}(t) < 1$, the limit of the numerator of Eq.(4.3) becomes infinite and the denominator draws closer to some positive value smaller than α as T_{psd} tends to infinity. Thus, the relation in Eq.(4.6) is established. This implies that the investigating search for a contact such that $pH_{sd}(\infty) < 1$ should be stopped sooner or later.

3. Since $H_{sd}(t)$ is continuous and differentiable in $\mathbf{T}^I$ by the assumption and the denominator of $E(\mathbf{T})$ given by Eq.(4.3) is positive in $\mathbf{T}^I$, $E(\mathbf{T})$ has no singular point and is continuous and differentiable in $\mathbf{T}^I$.

4. From Eq.(4.3), a unique singular point of $E(\mathbf{T})$ is $\{T_{psd} | T_{psd}=s$ for all $(p, s, d)\}$. As $\mathbf{T}$ approaches the singular point, $E(\mathbf{T})$ tends to infinity, and except for the singular point, $E(\mathbf{T})$ is continuous and has a finite positive value by Eq.(4.3). Therefore, $E(\mathbf{T})$ has a positive minimum value. □

We define a function $g_{psd}(t)$, $t \geqq s$,

(4.7) $$g_{psd}(t) = \frac{ph_{sd}(t)}{1-pH_{sd}(t)}.$$

It should be noted that $g_{psd}(t)$ is nonnegative and continuous with respect to t by the assumptions on $H_{sd}(t)$. Furthermore, we assume that $g_{psd}(t)$ is strictly decreasing in t, and $g_{psd}(\infty) = 0$. This assumption is reasonable and is usually satisfied in real-world applications, because $g_{psd}(t)$ means the marginal detection probability at t in the investigating search, and it usually becomes less efficient with the increase of t.

Here, we consider the following fractional programming problem and the associated problem. Let N and D be functions defined on a subset $\mathbf{S} \in \mathbf{R}^n$, N: $\mathbf{R}^n \to \mathbf{R}$ and D: $\mathbf{R}^n \to \mathbf{R}$. Problems A and B are defined by

Problem A: Minimize$_{\mathbf{x}}$ $\left[\frac{N(\mathbf{X})}{D(\mathbf{X})}\right]$ subject to $\mathbf{X} \in \mathbf{S}$, where $D(\mathbf{X}) > 0$ is assumed.

Problem B: Minimize$_{\mathbf{X}}$ $[N(\mathbf{X})-aD(\mathbf{X})]$ subject to $\mathbf{X} \in \mathbf{S}$,

where $a \in \mathbf{R}$ is a parameter. Let $\phi(a)$ be the optimal value function for Problem B,

$$\phi(a) = \inf\ [N(\mathbf{X})-aD(\mathbf{X}) \mid \mathbf{X} \in \mathbf{S}].$$

The following lemma is a fundamental theorem of fractional programming that we quote from a standard textbook with no proof.

Lemma 4.2. *If $\mathbf{X}^* \in \mathbf{S}$ is an optimal solution to Problem A, $\mathbf{X}^*$ is also an optimal solution to Problem B with $a = N(\mathbf{X}^*)/D(\mathbf{X}^*)$, and $\phi(a) = 0$ holds. On the contrary, if Problem B has an optimal solution $\mathbf{X}^* \in \mathbf{S}$ for an $a \in \mathbf{R}$, and if $\phi(a) = 0$, $\mathbf{X}^*$ is also an optimal solution to Problem A.*

Applying this lemma, we can derive the main theorem of our problem.

Theorem 4.3. *A necessary and sufficient condition for the investigating search plan $\mathbf{T}^*$ to be optimal is*

$$\text{(4.8a)} \qquad \text{if } g_{psd}(s) \leq \frac{1}{E(\mathbf{T}^*)}, \quad \text{then } T^*_{psd} = s,$$

and

$$\text{(4.8b)} \qquad \text{if } g_{psd}(s) > \frac{1}{E(\mathbf{T}^*)}, \quad \text{then } g_{psd}(T^*_{psd}) = \frac{1}{E(\mathbf{T}^*)}$$

for all (p, s, d).

Proof. Applying Lemma 4.2, our fractional programming problem defined by Eqs.(4.3) and (4.4) is led to a problem such that

$$\text{minimize}_{\mathbf{T}}\ G(\mathbf{T} \mid a) = 1 - a\alpha \int pH_{sd}(T_{psd})\,dB(p,s,d)$$

$$\text{(4.9)} \qquad + \alpha \int \Big[p\int_s^{T_{psd}} zh_{sd}(z)\,dz + (1-pH_{sd}(T_{psd}))\,T_{psd}\Big]\,dB(p,s,d),$$

subject to $T_{psd} \geq s$ and $\mathbf{T} \neq \{s\}$.

Let $\mathbf{T}^a$ be an optimal solution to Eq.(4.9) with a parameter a (>0). Since the objective function $G(\mathbf{T} \mid a)$ is separable in T_{psd}'s, $\mathbf{T}^a$ is obtained by minimizing $G_{psd}(T_{psd} \mid a)$ given by

$$G_{psd}(T_{psd} \mid a) = p\int_s^{T_{psd}} zh_{sd}(z)\,dz + (1-pH_{sd}(T_{psd}))\,T_{psd} - apH_{sd}(T_{psd})$$

for all T_{psd}. Since $H_{sd}(t)$ is a continuous and differentiable function of t by the assumptions (b) and (c) on $H_{sd}(t)$, $G_{psd}(T_{psd} \mid a)$ is differentiable with respect to T_{psd} and we have

$$\frac{dG_{psd}(T_{psd} \mid a)}{dT_{psd}} = ph_{sd}(T_{psd})\Big\{\frac{1}{g_{psd}(T_{psd})} - a\Big\}.$$

By the assumption of our model, $g_{psd}(T_{psd})$ is continuous and strictly decreasing from $g_{psd}(s)$ to zero when T_{psd} increases from s to infinity. Therefore, if and only if $g_{psd}(s) > 1/a$, $dG_{psd}(T_{psd}|a)/dT_{psd}$ changes its sign only once from negative to positive at T_{psd}, then $g_{psd}(T_{psd}) = 1/a$. Thus the solution to the equation $g_{psd}(T_{psd}) = 1/a$ gives the unique minimum of $G_{psd}(T_{psd}|a)$ if and only if $g_{psd}(s) > 1/a$. Meanwhile, if and only if $g_{psd}(s) \leqq 1/a$, $dG_{psd}(T_{psd}|a)/dT_{psd}$ is positive in $T_{psd} > s$, and therefore, $G_{psd}(T_{psd}|a)$ is minimum at the lower limit s. Thus we have

$$\text{(4.10a)} \qquad \text{iff} \quad g_{psd}(s) \leqq \frac{1}{a}, \quad \text{then} \quad T_{psd}{}^{a} = s,$$

$$\text{(4.10b)} \qquad \text{iff} \quad g_{psd}(s) > \frac{1}{a}, \quad \text{then} \quad g_{psd}(T_{psd}{}^{a}) = \frac{1}{a}.$$

On the other hand, if we set $a = E(\mathbf{T}^*)$ and $T_{psd} = T_{psd}{}^*$ in Eq. (4.9), we have $G(\mathbf{T}^*|a=E(\mathbf{T}^*)) = \phi(a=E(\mathbf{T}^*)) = 0$. Since $T_{psd}{}^*$ minimizing $G_{psd}(T_{psd}|a)$ for an a determined uniquely and $\phi(a=E(\mathbf{T}^*)) = 0$ as mentioned earlier, $\mathbf{T}^a$ given by Eqs. (4.10a) and (4.10b) with $a = E(\mathbf{T}^*)$ is optimal from Lemma 4.2. By setting $a = E(\mathbf{T}^*)$ and $T_{psd}{}^a = T_{psd}{}^*$ in Eqs. (4.10a) and (4.10b), we have Theorem 4.3. □

Since $g_{psd}(T_{psd})$ is assumed to be continuous and strictly decreasing in T_{psd}, $g_{psd}(T_{psd})$ admits an inverse function $g_{psd}{}^I(y)$. The function $g_{psd}{}^I(y)$ decreases from infinity to s when y increases from zero to $g_{psd}(s)$. We define $g_{psd}(y)^I = s$ if $y > g_{psd}(s)$. Hence, $g_{psd}{}^I(y)$ is defined as $g_{psd}{}^I: (0,\infty) \to [s,\infty)$. Using $g_{psd}{}^I$, we can state the theorem in another way.

Corollary 4.4. *A necessary and sufficient condition for the investigating search plan* $\mathbf{T}^*$ *to be optimal is*

$$\text{if } g_{psd}{}^I\left(\frac{1}{E(\mathbf{T}^*)}\right) = s, \quad \text{then } T_{psd}{}^* = s,$$

$$\text{if } g_{psd}{}^I\left(\frac{1}{E(\mathbf{T}^*)}\right) > s, \quad \text{then } T_{psd}{}^* = g_{psd}{}^I\left(\frac{1}{E(\mathbf{T}^*)}\right).$$

Here, we define two sets

$$\mathbf{I} = \{(p,s,d) \mid T_{psd}{}^* > s\},$$

$$\tilde{\mathbf{I}} = \{(p,s,d) \mid T_{psd}{}^* = s\}.$$

Set $\mathbf{I}$ is the set of contacts to be investigated and $\tilde{\mathbf{I}}$ is the set of contacts to be disregarded. Theorem 4.3 is stated in another way.

Corollary 4.5.

$$\textit{If} \quad p h_{sd}(s)\, E(\mathbf{T}^*) \leqq 1, \quad \textit{then } (p,s,d) \in \tilde{\mathbf{I}},$$

$$\textit{if} \quad p h_{sd}(s)\, E(\mathbf{T}^*) > 1, \quad \textit{then } (p,s,d) \in \mathbf{I}.$$

Proof. From Theorem 4.3, if $g_{psd}(s) \leqq 1/E(\mathbf{T}^*)$, then $T_{psd}{}^* = s$; namely, the contact (p,s,d) should be disregarded. Since $H_{sd}(s) = 0$,

$g_{psd}(s) = ph_{sd}(s)$ by the definition in Eq.(4.7). Therefore, the condition $g_{psd}(s) \leqq 1/E(\mathbf{T}^*)$ is rewritten as $ph_{sd}(s)E(\mathbf{T}^*) \leqq 1$. Similarly, if $ph_{sd}(s)E(\mathbf{T}^*) > 1$, $T_{psd}^* > s$ is established by Theorem 4.3. □

Remark 4.1. We consider the meaning of the conditions given by Eqs.(4.8a) and (4.8b). Substituting Eq.(4.7) into Eq.(4.8b), we have

$$\frac{ph_{sd}(T_{psd}^*)}{1-pH_{sd}(T_{psd}^*)} = \frac{1}{E(\mathbf{T}^*)}. \tag{4.11}$$

The left-hand side of this equation is the conditional detection probability per unit time at T_{psd}^* given that the investigating search for the contact (p,s,d) fails up to T_{psd}^*. In other words, this means the marginal detection probability when the investigating search is continued until T_{psd}^* and then it is prolonged by a unit time. Since $g_{psd}(t)$ is assumed to be strictly decreasing in t, this marginal detection probability decreases strictly to zero as T_{psd}^* increases. On the other hand, the right-hand side of Eq.(4.11), $1/E(\mathbf{T}^*)$, is the maximum value of the mean detection rate which is attained by employing the optimal investigating search plan $\mathbf{T}^*$. Therefore, the physical meaning of the conditions for the optimal investigating search plan, Theorem 4.3, is described as follows. If a contact is to be investigated, the stopping time should be determined in such a way that the marginal detection probability at the end of the investigating search is balanced to the mean detection rate in the overall search conducted optimally (Eq.(4.8b)). On the other hand, if a contact is to be disregarded, the reason is that the contact does not have a larger marginal detection probability than the mean detection rate of the optimal search (Eq.(4.8a)).

Usually, we must calculate $\mathbf{T}^*$ numerically, since the function g_{psd} is very complicated and the inverse function g_{psd}^I is not given in a closed form. The next theorem provides the base for the numerical calculation.

Theorem 4.6. *Let β be an arbitrary positive number and $\mathbf{T}(\beta) = \{T_{psd}(\beta)\}$ be an investigating search plan determined by $T_{psd}(\beta) = g_{psd}^I(\beta)$ for all (p,s,d). Then the following relation holds:*
If $\beta > 1/E(\mathbf{T}^*)$,

$$T_{psd}(\beta) \leqq T_{psd}^* \quad \text{and} \quad E(\{g_{psd}^I(\beta)\}) \geqq E(\mathbf{T}^*) > \frac{1}{\beta}, \tag{4.12a}$$

if $\beta < 1/E(\mathbf{T}^*)$,

$$T_{psd}(\beta) \geqq T_{psd}^* \quad \text{and} \quad E(\mathbf{T}^*) \leqq E(\{g_{psd}^I(\beta)\}) < \frac{1}{\beta}. \tag{4.12b}$$

Proof. We consider a contact such that $T_{psd}^* > s$ and define a function Z_{psd}: $(s,\infty)\to(-\infty,\infty)$ by

(4.13) $$Z_{psd}(T_{psd}) = \{\alpha \int pH_{sd}(T_{psd})dB(p,s,d)\}\{E(\mathbf{T}) - \frac{1}{g_{psd}(T_{psd})}\}.$$

The function $Z_{psd}(T_{psd})$ is continuous and differentiable with respect to T_{psd} ($>s$) by assumptions (b) and (c) on $H_{sd}(\cdot)$. We have

(4.14) $$\frac{dZ_{psd}(T_{psd})}{dT_{psd}} = \frac{\alpha \int pH_{sd}(T_{psd})dB(p,s,d)}{g_{psd}(T_{psd})^2} \times \frac{dg_{psd}(T_{psd})}{dT_{psd}}.$$

Since $g_{psd}(T_{psd})$ is assumed to be strictly decreasing in T_{psd}, $Z_{psd}(T_{psd})$ is strictly decreasing. Meanwhile, if $T_{psd}^* > s$, $g_{psd}(T_{psd}^*) = 1/E(\mathbf{T}^*)$ by Theorem 4.3. Thus $Z_{psd}(T_{psd}^*) = 0$ holds by Eq. (4.13), and if $T_{psd} <$ ($>$) T_{psd}^*, then $Z_{psd}(T_{psd}) >$ ($<$) 0. This implies

(4.15a) if $T_{psd} < T_{psd}^*$, then $\frac{1}{E(\mathbf{T})} < g_{psd}(T_{psd})$,

(4.15b) if $T_{psd} > T_{psd}^*$, then $\frac{1}{E(\mathbf{T})} > g_{psd}(T_{psd})$.

Here we consider the stopping time of the investigating search determined by $g_{psd}(T_{psd}) = \beta$ for a positive value β; $T_{psd}(\beta) = g_{psd}^{I}(\beta)$. Since $g_{psd}(T_{psd})$ is a decreasing function of T_{psd} ($>s$), the inverse function $g_{psd}^{I}(\beta)$ is also decreasing with respect to β. Thus if $\beta > 1/E(\mathbf{T}^*)$, $T_{psd}(\beta) = g_{psd}^{I}(\beta) < T_{psd}^*$ is valid, and $E(\{g_{psd}^{I}(\beta)\}) > 1/\beta$ is concluded from Eq. (4.15a). On the contrary, if $\beta < 1/E(\mathbf{T}^*)$, $T_{psd}(\beta) = g_{psd}^{I}(\beta) > T_{psd}^*$ holds, and in this case, $E(\mathrm{T}(\beta)) < 1/\beta$ from Eq. (4.15b). Thus, Theorem 4.6 is established. □

From this theorem, we can construct an algorithm for calculating the optimal stopping time and the expected time of detecting the target. This is the general flow of the calculation:

1. Set lattice points on the space $\{(p,s,d)\}$ for the calculation of the integration in Eq. (4.3).
2. Set a sufficiently small ε (>0) for the prescribed precision of the solution.
3. Set the index of the iteration $n = 1$.
4. Set a positive number $\beta_{n=1}$ such that $\beta_1 < \max_{(p,s,d)} g_{psd}(s)$.
5. Calculate the following value T_{psd}^n to all lattice points,

   if $g_{psd}(s) \leqq \beta_n$, then $T_{psd}^n = s$,
   if $g_{psd}(s) > \beta_n$, then $T_{psd}^n = g_{psd}^{I}(\beta_n)$.

6. Substitute $\mathbf{T}^n = \{T_{psd}^n\}$ into Eq. (4.3) and calculate $E(\mathbf{T}^n)$ numerically, e.g., by applying the Simpson rule.
7. If $|E(\mathbf{T}^n) - E(\mathbf{T}^{n-1})| \leqq \varepsilon$, stop the calculation and we have $\mathbf{T}^* \fallingdotseq \{T_{psd}^n\}$ and $E(\mathbf{T}^*) \fallingdotseq 1/\beta_n = E(\mathbf{T}^n)$.
8. If $|E(\mathbf{T}^n) - E(\mathbf{T}^{n-1})| > \varepsilon$, reset $n = n+1$, $\beta_{n+1} = 1/E(\mathbf{T}^n)$ and return to the fifth step.

In this calculation, if we start with β_1 such that $\beta_1 > 1/E(\mathbf{T}^*)$, the calculation proceeds $\mathbf{T}^1 < \mathbf{T}^2 > \cdots > \mathbf{T}^n \fallingdotseq \mathbf{T}^*$ by Eqs. (4.12a) and

(4.12b). On the other hand, if we start with $\beta_1 < 1/E(\mathbf{T}^*)$, then a sequence $\mathbf{T}^1 > \mathbf{T}^2 > \mathbf{T}^3 > \cdots > \mathbf{T}^n \fallingdotseq \mathbf{T}^*$ are produced by Eq. (4.12a). Thus $\{T_{psd}{}^*\}$ and $E(\mathbf{T}^*)$ are obtained approximately to within the prescribed accuracy.

4.2.2. Properties of the Optimal Investigating Search Plan

In this section, we examine the properties of the optimal investigating search plan given by Theorem 4.3.

Let a contact be represented by a point in the three-dimensional coordinate system (p, s, d). The detection function $h_{sd}(s)$ is here denoted simply by $h(s, d)$ and a surface specified by $ph(s, d)E(\mathbf{T}^*) = 1$ is called the critical contact surface. If $B(p, s, d)$ and $H_{sd}(t)$ are given, $E(\mathbf{T}^*)$ is determined and the critical contact surface is determined uniquely. A point on this surface is denoted by (p_0, s_0, d_0),

(4.16)
$$p_0 = \frac{1}{h(s_0, d_0)E(\mathbf{T}^*)}.$$

From Corollary 4.5, the space (p, s, d) is divided into two subspaces, $\mathbf{I}$ and $\tilde{\mathbf{I}}$, by the critical contact surface. If a contact belongs to the subspace $\tilde{\mathbf{I}}$; $ph(s, d)E(\mathbf{T}^*) \leqq 1$, the contact should be disregarded, and if it belongs to the subspace $\mathbf{I}$; $ph(s, d)E(\mathbf{T}^*) > 1$, it should be investigated. The critical contact surface defined by Eq. (4.16) has the following properties.

<u>Theorem 4.7.</u> 1. *p_0 is nondecreasing with respect to both s_0 and d_0.*

2. *If the efficiency of the search is improved, the subspace* $\mathbf{I}$ *diminishes in size.*

<u>Proof.</u> 1. Let the derivative of h with respect to x be $h_x{}'$. Since $h(s_0, d_0)$ is assumed to be nonincreasing and differentiable with respect to s_0, we have

$$\frac{\partial p_0}{\partial s_0} = -\frac{h_{s_0}{}'(s_0, d_0)}{E(\mathbf{T}^*)h^2(s_0, d_0)} \geq 0.$$

Therefore, p_0 is nondecreasing with respect to s_0. This means that if the time lag s increases, the investigating search should be limited to more reliable contacts.

Similarly, the derivative of p_0 with respect to d_0 is nonnegative, since $h(s_0, d_0)$ is assumed to be nonincreasing in d_0. Thus p_0 is nondecreasing in d_0. This implies that if the datum point error d increases, the investigating search should be limited to more reliable contacts.

2. As seen in Eq. (4.16), p_0 is inversely proportional to $E(\mathbf{T}^*)$.

Therefore, if $E(\mathbf{T}^*)$ decreases, then p_0 increases for a given (s_0, d_0) and the subspace $\mathbf{I}$ diminishes in size. This means that if the search becomes more efficient, for example in such a way that the quality and/or the rate of the contacts in the broad search are improved by using a highly efficient detection device, then the investigating search should be conducted by selecting more reliable contacts. □

From Theorem 4.7, the critical contact surface is schematically like that in Fig. 4.1.

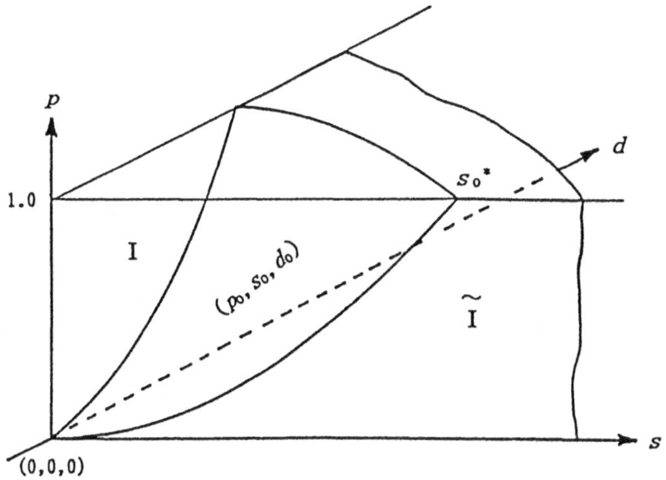

Figure 4.1. The critical contact surface.

Let us consider two arbitrary contacts, say contacts 1 and 2. The quality of contact and the stopping time are denoted by (p_i, s_i, d_i) and T_i, i=1,2, respectively, and $g_{psd}(t)$ is abbreviated as $g_i(t)$ for contact i, i=1,2. The optimal stopping times of the investigating search for these two contacts satisfy the following relation.

<u>Theorem 4.8.</u> *If contacts* 1 *and* 2 *are both to be investigated and* $g_1(t) < (>)\ g_2(t)$ *for all* t, *then the optimal stopping time of the investigating search for contact* 1 *is shorter (longer) than that for contact* 2, $T_1^* < (>)\ T_2^*$.

<u>Proof.</u> From Theorem 4.3, the optimal stopping time of the investigating search is determined by $g_i(T_i^*) = 1/E(\mathbf{T}^*)$ for all contacts $(p, s, d) \in \mathbf{I}$. Here we assume $T_1^* \geqq T_2^*$. Since $g_i(t)$ is strictly decreasing with respect to t, $g_1(T_1^*) \leqq g_1(T_2^*)$. Then $g_1(T_1^*) < g_2(T_2^*)$ is concluded by the assumption of the theorem. This contradicts the condition of the optimal stopping time, $g_i(T_i^*) = 1/E(\mathbf{T}^*)$. Thus,

$T_1^* < T_2^*$ is concluded. □

The following corollary is easily derived from Theorem 4.8.

Corollary 4.9. 1. *If* $p_1 < p_2$, $s_1 = s_2$ *and* $d_1 = d_2$, *then* $T_1^* < T_2^*$.

2. *If* $p_1 = p_2$, $s_1 < s_2$, $d_1 = d_2$, *and* $g_i(t)$ *is increasing decreasing) with respect to s, then* $T_1^* < (>)\ T_2^*$.

3. *If* $p_1 = p_2$, $s_1 = s_2$, $d_1 < d_2$, *and* $g_i(t)$ *is increasing (decreasing) with respect to d, then* $T_1^* < (>)\ T_2^*$.

Proof. From the definition of $g_i(t)$ given by Eq.(4.7), $\partial g_i(t)/\partial p_i > 0$ is easily derived. Therefore, if $p_1 < p_2$, $s_1 = s_2$, and $d_1 = d_2$, then $g_1(t) < g_2(t)$ for all t. Hence $T_1^* < T_2^*$ follows from Theorem 4.8.

If $g_i(t)$ is increasing with respect to s, $g_1(t) < g_2(t)$ holds for $s_1 < s_2$. Hence, $T_1^* < T_2^*$ is derived from Theorem 4.8 for Case 2. Case 3 is proved similarly. □

Remark 4.2. In our model, we assume that the searcher must not stop the evaluation of the contact quality. However, in real-world applications, if the evaluation of the contact is continued too long, the searcher should stop the evaluation to avoid the loss of search time. Let us consider the critical contact with $p_0 = 1$ and $d_0 = 0$, then we have the following relation from Eq.(4.16).

$$h(s_0, 0) = \frac{1}{E(\mathbf{T}^*)}. \tag{4.17}$$

Suppose Eq.(4.17) has a unique solution $s_0 = s_0^*$. Since the quality of contact $(p, s, d \mid s \geq s_0^*)$ is lower than that of contact $(1, s_0^*, 0)$, the contact $(p, s, d \mid s \geq s_0^*)$ belongs to $\tilde{\mathbf{I}}$. This implies that the evaluation of the contact should not be continued over s_0^*. Here, if the searcher is permitted to stop the evaluation of a contact, the search problem is considered to consist of three stages, i.e., the broad search, the evaluating search, and the investigating search. The stopping rule in the three-stage search is one of the problems to be investigated in the future.

4.3. Numerical Examples

Let us examine simple cases in which the possible contacts are made up of finite discrete points $\{(p, s, d)_i, i=1, \cdots, n\}$. In this model, $dB(p, s, d)$ is a probability mass function and all theorems obtained in the previous section are valid if integration with respect to $dB(p, s, d)$ is replaced by summation over i. In the following, we deal with four-point problems with $d = 0$,

$$\{(p, s, d{=}0)\} = \{(p_1, s_1, 0)_{i=1}, (p_1, s_2, 0)_2, (p_2, s_1, 0)_3, (p_2, s_2, 0)_4\}.$$

For simplicity, the triplet $(p_j, s_k, 0)$ will be abbreviated as (p_j, s_k) and subscripts ds in H_{sd}, g_{psd} and T_{psd} are omitted. The parameters p_j and s_k are assumed to be independent random variables:

$$dB(p_j, s_k) = \Pr(p_j)\Pr(s_k).$$

The function $H_s(t)$ is assumed to be

$$(4.18) \qquad H_s(t) = 1 - \exp\{-\frac{vW}{\pi u^2}(\frac{1}{s} - \frac{1}{t})\} \quad \text{for} \quad t \geqq s,$$
$$H_s(t) = 0 \quad \text{for} \quad t < s.$$

The detection function assumed by Eq.(4.18) corresponds to the following search situation. A searcher is searching for a target. At $t = 0$, the searcher got a contact; he got precise position of the target and then lost contact. He knows only the maximum speed u of the target and that the target will do its best to evade detection. Hence, it is natural to suppose the position of the target at τ is uniformly likely in the circular area with its center at the datum point and with radius $r = u\tau$. We assume that the searcher always searches the expanding circular area uniformly and randomly with speed v and effective sweep width W. The detection function of this datum search is derived easily and given by Eq.(4.18). It should be pointed out that the same function was also examined by Washburn (1981a).

This detection function satisfies assumption 5 of our model described earlier, and the function $g_{ps}(t)$ is easily obtained from Eqs.(4.7) and (4.18):

$$(4.19) \qquad g_{ps}(t) = \frac{pvW}{\pi u^2 t^2\{p+q \exp(A_s(t))\}},$$

where

$$A_s(t) = \frac{vW}{\pi u^2}(\frac{1}{s} - \frac{1}{t}) \quad \text{and} \quad q = 1 - p.$$

Since $g_{ps}(t)$ is continuous and strictly decreasing with respect to t in $t \geqq s$, the theorems presented in the previous section are applicable.

In the following, a standard case and its variants are analyzed to show numerical examples and to see the sensitivity of the optimal stopping time on the various parameters involved in the model. The standard case, Case 1, has parameter values such that

$$p_1 = 0.2, \quad p_2 = 0.7, \quad s_1 = 1 \text{ hr}, \quad s_2 = 3 \text{ hr},$$
$$\Pr(p_1) = \frac{2}{3}, \quad \Pr(p_2) = \frac{1}{3},$$
$$\Pr(s_1) = 0.7, \quad \Pr(s_2) = 0.3, \quad \alpha = 3,$$
$$u = 10 \text{ kt}, \quad vW = 750 \text{ nm}^2/\text{hr}.$$

1. Influence of s and p on T_{ps}^*

To see the influence of s and p on the optimal stopping time, we examine the next two cases.

Case 2: The time lag s_2 is twice as large as in Case 1, whereas the other parameters are kept the same as Case 1.

Case 3: Only p_2 is varied from 0.7 in Case 1 to 0.9.

The optimal stopping time T_{ps}^* is calculated by the algorithm given in the previous section with $\varepsilon = 0.001$ and is shown in Table 4.1. The convergence of the algorithm is very fast. The number of iteration of the calculation for these examples was not more than 5 if it started from an initial value $E(\mathbf{T}^1)$ longer than $E(\mathbf{T}^*)$ by about 10 (hours).

Table 4.1. The optimal stopping time T_{ps}^*.

Case Number		1	2	3
Contact		$p_2 = 0.7$ $s_2 = 3$	$p_2 = 0.7$ $s_2 = 6$	$p_2 = 0.9$ $s_2 = 3$
No.	(p_j, s_k)			
1	(0.2, 1)	1.65(0.65)	1.83(0.83)	1.55(0.55)
2	(0.2, s_2)	3.00 (0)	6.00 (0)	3.00 (0)
3	(p_2, 1)	3.11(2.11)	3.56(2.56)	3.86(2.86)
4	(p_2, s_2)	4.44(1.44)	6.00 (0)	4.63(1.63)
$E(\mathbf{T}^*)$		12.84	17.91	10.31

Note: ( ) = net search time for investigation.

In Fig. 4.2, the four possible contacts $\{(p_j, s_k)_i\}$ are plotted together with the curve of the critical contact which is a special case of Eq. (4.16) with $d = 0$:

$$p_0 = \frac{\pi u^2}{vWE(\mathbf{T}^*)} s_0^2.$$

In Case 1, only contact 2 should be disregarded as shown in Table 4.1. This assertion is also confirmed from the fact that point 2 in Fig. 4.2, Case 1, is below the curve of the critical contact.

Case 2 is characterized by a longer time lag s_2. As shown in Fig. 4.1, both points 2 and 4 have abscissa 6 instead of 3 in Case 1 and are below the curve of the critical contact; we should disregard contacts 2 and 4. This result is also seen in Table 4.1. The table also shows that in Case 2 the investigating search for contacts 1 and 3 should be longer than the search in Case 1.

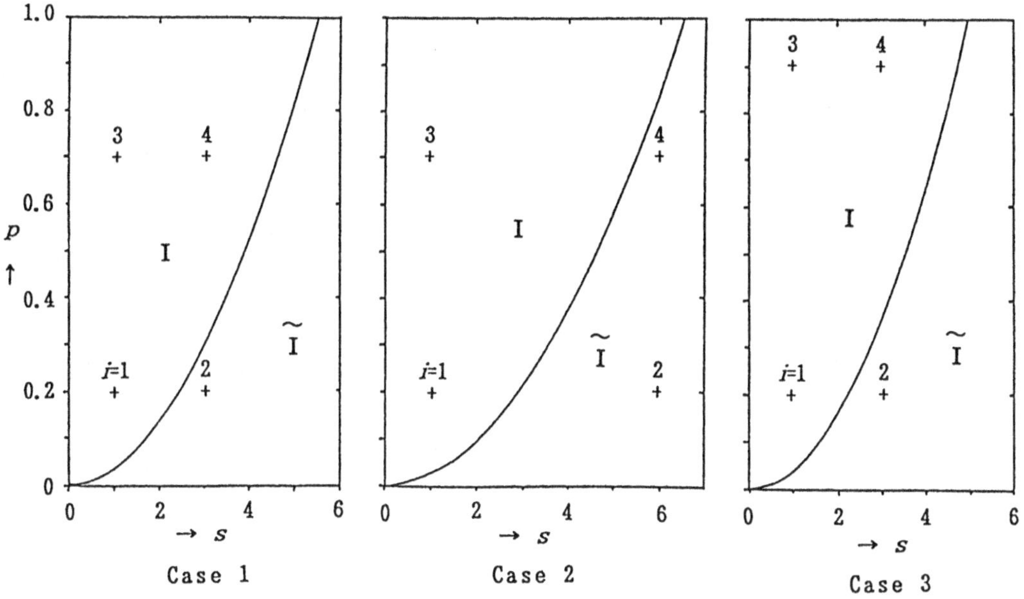

Figure 4.2. Contacts in **I** and $\tilde{\mathbf{I}}$.

Case 3 is similar to Case 1 in that only contact 2 should be disregarded. It should be noted that in Case 3 the investigating search times for contacts 3 and 4 become longer and the time for contact 1 is shortened compared with those in Case 1. In each case, the property of the optimal stopping time stated in Corollary 4.9-1 is confirmed by comparing $T_{ps}{}^*$ for contacts 1 and 2 with that for contacts 3 and 4 respectively. Also the property stated in Corollary 4.9-2 is confirmed by comparing contacts 1 and 3 with contacts 2 and 4, respectively, since the function $g_{ps}(t)$ given by (4.19) is an increasing function of s.

2. Influence of α and $B(p, s)$

We examine the next three cases to see the influence of the rate α of the contacts and the probability distribution $B(p, s) = \Pr(p)\Pr(s)$ on the optimal stopping plan $T_{ps}{}^*$.

Case 4 : α is one-tenth of the rate in Case 1.
Case 5 : $(\Pr(p_1), \Pr(p_2))$ is varied from (2/3, 1/3) in Case 1 to (1/3, 2/3).
Case 6 : $(\Pr(s_1), \Pr(s_2))$ is varied from (0.7, 0.3) in Case 1 to (0.3, 0.7).

The optimal stopping times for these cases are shown in Table 4.2. Table 4.2 shows that all the contacts should be investigated in Case 4, and the search time is much longer than in Case 1. This is because Case 4 is in contrast with Case 1 in its infrequent occurrence of incoming

Table 4.2. The optimal stopping time T_{ps}^*.

Case Number		1	4	5	6
Contact			α = 0.3	$(\Pr(p_j))$ =(1/3, 2/3)	$(\Pr(s_k))$ =(0.3, 0.7)
No.	(p_j, s_k)				
1	(0.2, 1)	1.65(0.65)	2.07(1.07)	1.48(0.48)	2.06(1.06)
2	(0.2, 3)	3.00 (0)	3.42(0.42)	3.00 (0)	3.39(0.39)
3	(0.7, 1)	3.11(2.11)	4.18(3.18)	2.69(1.69)	4.14(3.14)
4	(0.7, 3)	4.44(1.44)	6.21(3.21)	3.74(0.74)	6.14(3.14)
$E(\mathbf{T}^*)$		12.84	26.57	8.82	25.98

Note: ( ) = net search time for investigation.

contacts; the rate is only one-tenth of that in Case 1. We see that the search times in the column for Case 5 are shorter than the corresponding values for Case 1. This would be a natural consequence of the frequent occurrence of high-quality contacts in Case 5 compared with Case 1. In Case 6, most of the contacts have long time lags, in contrast to Case 1. The table shows full investigation of all the contacts is needed.

In Theorem 4.7-2, we pointed out that the subspace I diminishes in size if the quality of contact is improved and/or the rate of incoming contacts increases. Fig. 4.3 illustrates this assertion.

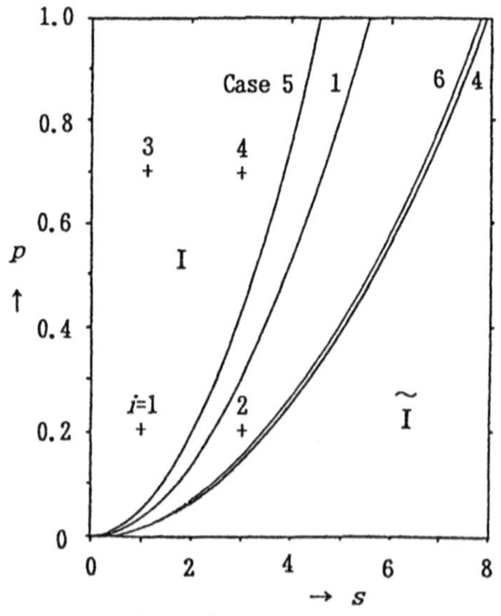

Figure 4.3. Critical contact surface.

3. Effect of Variation in u and vW

Finally, the following two cases are examined to see the effect of variation in the target's speed u and the sweep rate vW of the searcher on the optimal stopping plan.

Case 7: The target's speed u is three times as large as u in Case 1.

Case 8: The sweep rate of the investigating search vW = 3000 (nm^2/hr) is 4 times as large as the value in Case 1.

The optimal stopping times for the investigating search for Cases 7 and 8 are shown in Table 4.3.

Table 4.3. The optimal stopping Time T_{ps}^*.

Case Number		1	7	8
Contact		u = 10 vW = 750	u = 30	vW = 3000
No.	(p_j, s_k)			
1	(0.2, 1)	1.65(0.65)	1.87(0.87)	1.32(0.32)
2	(0.2, 3)	3.00 (0)	3.00 (0)	3.34(0.34)
3	(0.7, 1)	3.11(2.11)	3.57(2.57)	1.32(0.32)
4	(0.7, 3)	4.44(1.44)	3.67(0.67)	5.23(2.23)
$E(\mathbf{T}^*)$		12.84	73.02	7.62

Note: ( ) = net search time for investigation.

The column for Case 7 shows that if the target speed is high, the investigating search effort should be concentrated on the contacts with short time lags. The optimal investigation times for contacts 1 and 3 are longer than the corresponding values in Case 1, whereas the investigation time for contact 4 is less than half of the value in Case 1.

In Case 8, the investigating search for the contact with short time lags (contacts 1 and 3) are stopped soon and the contacts with long time lags (contacts 2 and 4) are investigated thoroughly. The reason is as follows. Since the sweep rate is considerably large, the contacts with short time lags will be detected soon if the contact is true. Therefore, the contacts with short time lags need not be investigated for long. In this case, even the contacts with long time lags are possibly detected in the investigating search if plenty of time is applied. Hence, the stopping times for contacts 2 and 4 are long.

<u>Remark 4.3.</u> Here we consider the case of the exponential detection

function; let us assume

(4.20) $$H_{sd}(t) = 1 - \exp(-\gamma(d)(t-s)),$$
$$h_{sd}(t) = \gamma(d)\exp(-\gamma(d)(t-s)),$$

where $\gamma(d)$ is a constant specified by d. Substituting Eq.(4.20) into Eq.(4.7), for $t \geqq s$, we have

(4.21) $$g_{psd}(t) = \frac{p\,\gamma(d)}{p + q\exp(\gamma(d)(t-s))},$$
$$q = 1 - p.$$

Since $g_{psd}(t)$ given by Eq.(4.21) is continuous and strictly decreasing in t, the assumption of our model is satisfied and the theorems can be applied. Substituting Eq.(4.20) into Eq.(4.16), we have the critical contact surface:

(4.22) $$p_0 = \frac{1}{E(\mathbf{T}^*)\gamma(d_0)}.$$

In this relation, $E(\mathbf{T}^*)$ is determined uniquely if $B(p, s, d)$ is given, and therefore, the critical contact (p_0, s_0, d_0) given by Eq.(4.22) does not depend on time lag s_0. The reason why p_0 is independent of s_0 can be explained as follows. Since the conditional detection function Eq.(4.20) is exponential, the investigation process has the Markov property and the conditional detection rate in the investigating search is constant at any time. In this case, detection of the target in the investigating search does not depend on time lag s, but on search time $(t-s)$. This is why the critical contact does not depend on time lag s_0.

From the relation in Eq.(4.21), we can derive the inverse function $g_{psd}'(\cdot)$ as

$$g_{psd}'(x) = s + \frac{1}{\gamma(d)}\log\frac{p}{q}\left(\frac{\gamma(d)-x}{x}\right).$$

However, even in this case, it is difficult to derive the optimal investigating search plan $\mathbf{T}^*$ in a closed form.

CHAPTER 5

Forestalling Detection in Two-Sided Search

5.1. Introduction

In this chapter, we will investigate a forestalling detection problem in a two-sided search situation. At present, studies of the two-sided search problem are classified into two types, the hide-and-search problem and the evasion-and-search problem. In the hide-and-search problem, the target hides itself, selecting a point in the target space and stays there inactive during the search. A searcher wants to find the target efficiently, whereas the target does want not to be found. Usually, the game theoretical approach is taken and the optimal strategies for both sides are sought. The works of Neuts (1963), Johnson (1964), Danskin (1968), and Gittins and Roberts (1979) are the studies of this category. In the evasion-and-search problem, the target is assumed to be able to evade the searcher, observing the searcher's current position during the search. The problem is formulated as a sequential multistage search game. Norris (1962), Sakaguchi (1973), Washburn (1980b), and Nakai (1986a) investigated this problem. Nakai (1986b) considered a search game in which two searchers compete with each other in detecting a stationary and unconscious target.

There is another type of search in which both the searcher and the target are searching for each other, but somehow there is no paper dealing with the optimal search plan of this problem. In this chapter, we will investigate the characteristics of the search process (not the optimal policy) of this type, particularly the forestalling detection probabilities of both sides in the broad search and the barrier search.

The forestalling detection problem was dealt with first by Koopman (1946) in section 2.8 of "Search and Screening." In his book, he considered two problems: the forestalling detection probability of the target in the two-sided search with an exponential detection function and the forestalling sweep width of the searcher with an inverse cube law of detection. The forestalling detection problem in the broad search was dealt with by Giammo (1963). He investigated a search model in which a target and a searcher search for each other with intermittent moves in a definite area and derived the probability of forestalling detection of the target by the searcher.

As for the two-sided barrier search problem, several studies are reported: Washburn (1974), Ruckle et al. (1976), Ruckle (1981), Ruckle and Reay (1981), and Baston and Bostock (1987). All of these studies investigate the hide-and-search problem in the barrier search.

In this chapter, two problems of the forestalling detection in two-sided search are considered. First, in Section 5.2, we deal with a broad search problem that takes forestalling detection by the target into account and investigate the search process by a Markov chain model. This section is based on Iida (1985). Then, in Section 5.3, we consider a two-sided barrier search problem and derive the forestalling detection probabilities for both sides. This section is taken from Iida (1988b).

5.2. Forestalling Detection in Broad Search

The search process dealt with in this section is as follows.

1. Suppose a searcher is searching for a target in a region with area A that is homogeneous with respect to the detectability of detection device. This region is referred to as the target space.

2. The target arrives in the target space according to an exponential probability distribution with mean $1/\alpha_A$, and hides itself, selecting a point in the target space randomly and uniformly. The duration of the target's stay in the target space is an exponentially distributed random variable with mean $1/\alpha_D$. Once the target leaves the target space, it never comes back and the searcher cannot obtain any information of the target thereafter.

3. At the starting time of the search, the searcher is assumed to be able to estimate three probabilities: the target has not yet appeared, is staying in the target space, and has already left the target space. The probabilities are denoted by $p_1{}^0$, $p_2{}^0$, and $p_5{}^0$, respectively, $p_1{}^0 + p_2{}^0 + p_5{}^0 = 1$.

4. The searcher and the target search for each other continuously with detection devices characterized by instantaneous detection probability rates $b_S(r)$ and $b_T(r)$, respectively, where r is the distance between the searcher and the target.

5. If the target forestalls the searcher in detection, it immediately reacts in either of two ways, as a friend or as a foe. The instantaneous detection rate $b_S(r)$ and $b_T(r)$ are assumed to change to $b_S{}^0(r)$ and $b_T{}^0(r)$, respectively. If the target is a friend, such as a victim in the ocean waiting for rescue, the detectability of the detection device of the searcher is improved by the target's reaction; if it is a hostile target, the detectability is decreased by the target's alertness. Usually the detectability of the detection device

of the target is improved by its alertness. Here we assume the reaction or alertness of the target to the searcher is only for a while. The duration of the reaction is assumed to be an exponential random variable with mean $1/\alpha_R$, and after that the detectabilities of the detection devices come back to $b_S(r)$ and $b_T(r)$, respectively.

6. The searcher searches for the target with speed v in the target space. The search pattern of the searcher is assumed to be random in the sense that the path can be thought of as having its different (not too near) portions placed independently of one another. Any portion is assumed to be small relative to the total length of path but is decidedly larger than the range of possible detection and each portion of the path is considered an approximately straight line. This search pattern is called random search.

This search process has two distinctive features. The target dealt with here is an appearing and disappearing target and the detectability of the detection device changes when the target forestalls the searcher in detection. In the following, we formulate the search process as a continuous-time Markov chain model and derive the several characteristics of the search process, such as the probability of the process being in a specified state at t, the expected number of the target entering a specified state until t, the expected detection time, or the expected leaving time of the target from the target space.

We can categorize the state of the search process into five states defined by the following.

State 1 (target not yet appeared): In this state, the target has not appeared in the target space and no detection is expected; the searcher is wasting his searching time.

State 2 (target appeared): The searcher and the target are trying to find each other in this state.

State 3 (target on the alert): The target has forestalled the searcher in detection and is now in an alert situation. The detection laws of the detection devices of the searcher and target are specified by $b_S{}^0(r)$ and $b_T{}^0(r)$, respectively.

State 4 (target detected): The searcher detected the target successfully and the process came to an end. This state is one of the absorbing states.

State 5 (target disappeared): The target left the target space with no return. In this state, detection never occurs. It is also an absorbing state.

The process goes from one state to another when a specified event occurs in the former state. The transitions of the process among states are shown in Fig. 5.1.

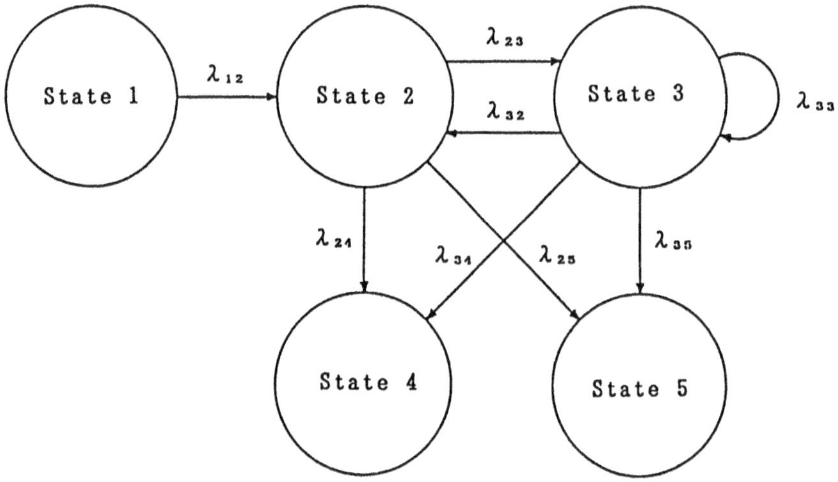

Figure 5.1. Transition of states.

The formulation of the search process dealt with here is accomplished by specifying the matrix of the transition rate. Let λ_{ij} be the conditional transition rate to state j given that the process is now in state i. The transition rates λ_{12}, λ_{25}, λ_{32}, and λ_{35} in Fig. 5.1 are determined directly from the assumptions of the model.

(5.1)
$$\lambda_{12} = \alpha_A, \qquad \lambda_{25} = \alpha_D,$$
$$\lambda_{32} = \alpha_R, \qquad \lambda_{35} = \alpha_D.$$

The other transition rates λ_{23}, λ_{24}, λ_{33}, and λ_{34} are calculated by considering the relevant events in the search.

Suppose the process is now in State 2; that is, the searcher and the target are in the target space A and are trying to find each other. Let R be the maximum range of the detection devices of both sides and assume $R^2 \ll A$. An encounter is defined as an event in which the distance between the searcher and the target goes down and gets to R. Since we assume the random search and uniform distribution of the target in the target space, the time to an encounter is an exponentially distributed random variable with rate $2Rv/A$. We denote it as λ_E, $\lambda_E = 2Rv/A$.

There are three possibilities following an encounter. The first is the case where both sides miss the chance of detecting the other sides. In this case the state is left unchanged. In the second case, the target forestalls the searcher and the encounter ends leaving the unconscious searcher. The state becomes State 3 in this case. The last case is the detection case. The searcher detects the target during the encounter, and the state becomes the absorbing state 4. We consider the transition to occur during an encounter at the instant of occurrence of the encounter since the duration is much smaller than the overall search time. Let the conditional probabilities that the state becomes to 3 and 4 when an encounter occurs be μ_{23} and μ_{24}, respectively. Then the time to a transition to State 3 or 4 is an exponentially distributed random variable with rate λ_{23} or λ_{24}, where

(5.2) $$\lambda_{2j} = \lambda_E \mu_{2j}, \quad j = 3, 4.$$

In order to derive expressions for μ_{2j}, we consider the following submodel.

Suppose the state was in State 2 and an encounter with lateral range x occurred at some time point. Without any loss of generality, we take the closest point of approach as the position corresponding to the origin of time, $t = 0$. In the following, let subscripts S and T stand for the searcher and target, respectively; we use k in place of (S or T). Let $p_k(t|x)$ be the conditional probability density of the forestalling detection by k at t conditional on the fact that the encounter is with lateral range x, and let $r(t|x)$ be the distance between the searcher and target at t. Then, we have

$$p_k(t|x) = b_k(r(t|x)) \exp[-\int_{-\infty}^{t} \{b_S(r(z|x)) + b_T(r(z|x))\} dz],$$

where the lower limit of the integral should be the time when the encounter begins, since the detection does not occur outside the range R, however, it can be replaced by $-\infty$ without any change in the value of the integral. By integrating $p_k(t|x)$ with respect to t, we obtain the conditional forestalling probability by k during the encounter,

$$FP_k(x) = \int_{-\infty}^{\infty} p_k(t|x)\, dt.$$

In this way, the searcher can detect the target by forestalling. But another way of detection is also possible ; namely, the target forestalls and gets into alert status, but during the rest of the encounter the searcher detects the target. A conditional detection probability of this type is given by the formula,

$$AP_S(x) = \int_{-\infty}^{\infty} p_T(t|x)\{1-\exp(-\int_{t}^{\infty} b_S{}^0(r(z|x))\, dz)\}\, dt.$$

Therefore, the conditional detection probability $PP_S(x)$ by the searcher is given by

$$PP_S(x) = FP_S(x) + AP_S(x),$$

which corresponds to the transition to State 4. Another change of state results from the one-way detection by the target; namely, the target forestalls and gets into alert status, and no detection by the searcher takes place during the rest of the encounter. The conditional probability of this event is given by

$$PP_T(x) = FP_T(x) - AP_S(x) = \int_{-\infty}^{\infty} p_T(t|x)\exp(-\int_t^{\infty} b_S{}^0(r(z|x))dz)dt.$$

Since the lateral range x is uniformly distributed over $[-R,R]$, the unconditional probabilities are given by

(5.3) $$PP_k = \frac{1}{2R}\int_{-R}^{R} PP_k(x)dx.$$

Since $\mu_{23} = PP_T$ and $\mu_{24} = PP_S$, λ_{2j}, j=3, 4, are obtained from Eqs. (5.2) and (5.3), and the relation $\lambda_E = 2Rv/A$.

(5.4) $$\lambda_{23} = \frac{v}{A}\int_{-\infty}^{\infty}\int_{-\infty}^{\infty}\Lambda_{23}(t,x)dtdx,$$

$$\lambda_{24} = \frac{v}{A}\int_{-\infty}^{\infty}\int_{-\infty}^{\infty}\Lambda_{24}(t,x)dtdx,$$

where

(5.5) $$\Lambda_{23}(t,x) = b_T(r(t|x))\exp[-\int_{-\infty}^{t}\{b_S(r(z|x))+b_T(r(z|x))\}dz - \int_t^{\infty} b_S{}^0(r(z|x))dz],$$

$$\Lambda_{24}(t,x) = [b_S(r(t|x)) + b_T(r(t|x))\{1-\exp(-\int_t^{\infty} b_S{}^0(r(z|x))dz)\}] \times \exp[-\int_{-\infty}^{t}\{b_S(r(z|x))+b_T(r(z|x))\}dz].$$

The transition rates λ_{33} and λ_{34} are derived in a similar way. Suppose the process is now in State 3 and an encounter with lateral range x happens to occur. In deriving μs, the expressions $FP_k(x)$, $AP_k(x)$, $PP_k(x)$ and PP_k obtained earlier are valid in this case if $b_k(r)$ is interchanged by $b_k{}^0(r)$, k = S or T. Therefore, we have

(5.6) $$\lambda_{33} = \frac{v}{A}\int_{-\infty}^{\infty}\int_{-\infty}^{\infty}\Lambda_{33}(t,x)dtdx,$$

$$\lambda_{34} = \frac{v}{A}\int_{-\infty}^{\infty}\int_{-\infty}^{\infty}\Lambda_{34}(t,x)dtdx,$$

where

(5.7) $$\Lambda_{33}(t,x) = b_T{}^0(r(t|x))\exp[-\int_{-\infty}^{t}\{b_S{}^0(r(z|x))+b_T{}^0(r(z|x))\}dz - \int_t^{\infty} b_S{}^0(r(z|x))dz],$$

$$\Lambda_{34}(t,x) = [b_S{}^0(r(t|x)) + b_T{}^0(r(t|x))\{1-\exp(-\int_t^{\infty} b_S{}^0(r(t|x))dz)\}] \times \exp[-\int_{-\infty}^{t}\{b_S{}^0(r(t|x))+b_T{}^0(r(t|x))\}dz].$$

Therefore, all the transition rates λ_{ij} are obtained by Eqs. (5.1) and (5.4)~(5.7).

Thus far, we have dealt with general instantaneous detection functions $b_k(r)$ and $b_k{}^0(r)$, $k = S, T$. Now we will consider the case of inverse cube law of detection.

$$b_k(r) = \frac{m_k}{r^3}, \qquad k = S, T,$$

$$b_S{}^0(r) = \frac{\beta\, m_S}{r^3},$$

$$b_T{}^0(r) = \frac{\gamma\, m_T}{r^3}.$$

In these formulas, if the target is a friend, then $\beta > 1$, otherwise, $0 \leqq \beta \leqq 1$. As mentioned before, usually $\gamma > 1$. Substituting these functions into Eqs.(5.4)～(5.7) and rewriting m_k by the sweep width $W_k = 2(2\pi\, m_k/v)^{1/2}$ in a one-sided search, we have λ_{2j} and λ_{3j} as follows.

$$\lambda_{23} = \frac{\alpha^2}{\sqrt{\alpha^2+1}+\sqrt{\beta}} \times \frac{vW_S}{A},$$

$$\lambda_{24} = \frac{\sqrt{\beta(\alpha^2+1)}+1}{\sqrt{\alpha^2+1}+\sqrt{\beta}} \times \frac{vW_S}{A},$$

$$\lambda_{33} = (\sqrt{\alpha^2\gamma+\beta}-\sqrt{\beta}) \times \frac{vW_S}{A},$$

$$\lambda_{34} = \sqrt{\beta} \times \frac{vW_S}{A},$$

where $\alpha = W_T/W_S$.

Let T_i be the staying time in State i and $F_i(t)$ be the distribution function (tail) of T_i, $F_i(t) = \Pr(T_i > t)$. Then $F_i(t)$ satisfies the following relation.

$$F_i(t) = \exp(-(\lambda_E+\lambda_{i5})t) + \int_0^t \lambda_E(1-\mu_{i3}-\mu_{i4})F_i(t-z)\exp(-(\lambda_E+\lambda_{i5})z)\,dz.$$

This equation is easily solved by using the Laplace transform and we have

(5.8) $$F_i(t) = \exp(-\Sigma_j \lambda_{ij} t).$$

Therefore, the transition probability density functions $g_{ij}(t)$ are obtained by

(5.9) $$g_{ij}(t) = \lambda_{ij}F_i(t) = \lambda_{ij}\exp(-\Sigma_j \lambda_{ij} t).$$

Our continuous-time Markov chain model is determined completely by the transition probability density functions $g_{ij}(t)$ given by Eq.(5.9). The following relations are well known,

$$g_{ij}(t) = p_{ij}f_{ij}(t),$$

$$p_{ij} = \int_0^\infty g_{ij}(t)\,dt,$$

$$f_{ij}(t) = \frac{g_{ij}(t)}{\int_0^\infty g_{ij}(t)\,dt} = -\frac{dF_i(t)}{dt}.$$

Defining $\lambda_i = \Sigma_j \lambda_{ij}$, for $i = 1,2,3$, we have

$$p_{ij} = \frac{\lambda_{ij}}{\lambda_i},$$
$$f_{ij}(t) = \lambda_i \exp(-\lambda_i t) \equiv f_i(t).$$

In the following, we derive the renewal density function, the expected number of events in a state until t, the probability distribution of states at t, and the expected time until absorption.

1. The Renewal Density Function

Suppose the process is in State i at $t = 0$. The renewal density functions $h_{ij}(t)$ are defined by

$$h_{ij}(t) = \lim_{\Delta t \to 0} \frac{1}{\Delta t} \Pr\left[\begin{array}{l} \text{the process goes into} \\ \text{State } j \text{ in } (t, t+\Delta t] \end{array} \middle| \begin{array}{l} \text{State } i \\ \text{at } t=0. \end{array} \right]$$

The functions $h_{ij}(t)$ satisfy the following relations.

$$h_{ij}(t) = g_{ij}(t) + \Sigma_k \int_0^t h_{ik}(t-z) g_{kj}(z)\,dz. \tag{5.10}$$

Here we denote the Laplace transform of a function $Y(t)$ by $Y^*(s)$. On taking Laplace transforms of Eq. (5.10), we have simultaneous linear equations of $h_{ij}^*(s)$. Then the $h_{ij}^*(s)$s are solved as

$$[h_{ij}^*(s)] = [g_{ij}^*(s)][I-[g_{ij}^*(s)]]^{-1}, \tag{5.11}$$

where [ ] is a 5 X 5 matrix, $[\ ]^{-1}$ is an inverse matrix, and I is a unit matrix. The renewal density function $h_{ij}(t)$ is obtained by inversely transforming $h_{ij}^*(s)$ given by Eq. (5.11).

If the initial state is not certain and the searcher knows only a probability distribution,

$$\mathbf{p}^0 = \{p_1^0,\ p_2^0,\ 0,\ 0,\ p_5^0\},$$

the renewal probability density functions are given by a 1 X 5 vector $\mathbf{h}(t)$ defined by

$$\mathbf{h}(t) = \{\Sigma_i p_i^0 h_{ij}(t)\}.$$

In the following, we derive $\mathbf{h}(t)$ in a closed form. The function $\mathbf{h}(t)$ obtained by the Laplace inverse transform is classified by the roots of the following equation, the left-hand side of which is the denominator of $h_{ij}^*(s)$ given by Eq. (5.11),

$$(s+\lambda_1)\{s^2+(\lambda_2+\lambda_3(1-p_{33}))s+\lambda_2\lambda_3(1-p_{33}-p_{23}p_{32})\} = 0.$$

It is easily confirmed that the quadratic equation in the braces has two negative roots, and so the preceding equation has three negative roots $(-a_1, -a_2, -a_3)$,

$$a_1 = \lambda_1,$$
$$a_2 = \frac{\lambda_2+\lambda_3(1-p_{33})-D}{2},$$
$$a_3 = \frac{\lambda_2+\lambda_3(1-p_{33})+D}{2},$$
$$D = \sqrt{\{\lambda_2-\lambda_3(1-p_{33})\}^2+4\lambda_{23}\lambda_{32}}.$$

Since $a_2 \neq a_3$, the following two cases are possible.

Case 1: $a_1 \neq a_2$ and $a_1 \neq a_3$.

(5.12a) $$\mathbf{h}(t) = \sum_{n=1}^{3} \mathbf{K}(-a_n)\exp(-a_n t),$$

where

$$\mathbf{K}(x) = \frac{\mathbf{B}_1 x^2+\mathbf{B}_2 x+\mathbf{B}_3}{3x^2+2(a_1+a_2+a_3)x+a_1 a_2+a_2 a_3+a_1 a_3}.$$

Case 2: $a_1 = a_2$ or $a_1 = a_3$. The double root is denoted by a_0, and the other root by a_*.

(5.12b) $$\mathbf{h}(t) = \frac{\mathbf{B}_1 a_*^2-\mathbf{B}_2 a_*+\mathbf{B}_3}{(a_*-a_0)^2}\exp(-a_* t)+\left[\frac{\mathbf{B}_1 a_0(a_0-2a_*)+\mathbf{B}_2 a_*-\mathbf{B}_3}{(a_*-a_0)^2} + \frac{\mathbf{B}_1 a_0^2-\mathbf{B}_2 a_0+\mathbf{B}_3}{a_*-a_0}\,t\right]\exp(-a_0 t).$$

In these equations, the $\mathbf{B}_j$s are 1 X 5 vectors and are given as follows.

$$\mathbf{B}_1 = \{\,0,\ p_1{}^0\lambda_1,\ p_2{}^0\lambda_{23},\ p_2{}^0\lambda_{24},\ p_2{}^0\lambda_{25}\,\},$$
$$\mathbf{B}_2 = \{\,0,\ p_1{}^0\lambda_1 b_4+p_2{}^0\lambda_{23}\lambda_{32},\ \lambda_{23}(p_1{}^0\lambda_1+p_2{}^0(\lambda_1+\lambda_3)),\ p_1{}^0\lambda_1\lambda_{24}+p_2{}^0(b_1+\lambda_1\lambda_{24}),\ p_1{}^0\lambda_1\lambda_{25}+p_2{}^0(b_2+\lambda_1\lambda_{25})\},$$
$$\mathbf{B}_3 = \{\,0,\ p_1{}^0\lambda_1\lambda_2(\lambda_3-\lambda_{33})+p_2{}^0\lambda_1\lambda_{23}\lambda_{32},\ (p_1{}^0+p_2{}^0)\lambda_1\lambda_3\lambda_{23},\ (p_1{}^0+p_2{}^0)\lambda_1 b_1,\ (p_1{}^0+p_2{}^0)\lambda_1 b_2\,\},$$

where

$$b_1 = \lambda_2\lambda_3\{p_{24}(1-p_{33})+p_{23}p_{34}\},$$
$$b_2 = \lambda_2\lambda_3\{p_{25}(1-p_{33})+p_{23}p_{35}\},$$
$$b_3 = \lambda_2\lambda_3\{1-p_{33}+p_{23}p_{32}\},\ (b_3 \text{ is used later}),$$
$$b_4 = \lambda_2+\lambda_3(1-p_{33}).$$

2. The Expected Number of Transitions

The expected number of transitions $H_{ij}(t)$ to State j until t given the process being in State i at $t = 0$ is:

$$H_{ij}(t) = \int_0^t h_{ij}(z)\,dz.$$
$$\mathbf{H}(t) = \{\Sigma_i p_i{}^0 H_{ij}(t)\}.$$

Substituting Eqs.(5.12a) and (5.12b) into this relation, we obtain

Case 1:

(5.13a) $$\mathbf{H}(t) = \sum_{n=1}^{3}\frac{\mathbf{K}(-a_n)}{a_n}\{1-\exp(-a_n t)\}.$$

Case 2:

(5.13b)
$$H(t) = \frac{B_1 a_*^2 - B_2 a_* + B_3}{a_*(a_* - a_0)^2}\{1-\exp(-a_* t)\} - \frac{B_1 a_* a_0^2 - B_2 a_0^2 - B_3(a_* - 2a_0)}{a_0^2(a_* - a_0)^2}\{1-\exp(-a_0 t)\} - \frac{B_1 a_0^2 - B_2 a_0 + B_3}{a_0(a_* - a_0)} t \exp(-a_0 t).$$

Since $a_n > 0$, the expected number of transitions to State j before absorption is given by $\lim_{t\to\infty} H(t)$.

$$\lim_{t\to\infty} H(t) = H(\infty) = \frac{B_3}{\lambda_1 b_3}.$$

3. The State Distribution

The probability $P_{ij}(t)$ of the process being in State j at t given the initial state i satisfies the following relation,

$$P_{ij}(t) = \delta_{ij} F_i(t) + \int_0^t h_{ij}(t-z) F_j(z)\, dz,$$
$$\delta_{ij} = 1 \quad \text{if } i = j, \text{ and } \delta_{ij} = 0 \quad \text{if } i \neq j,$$

where $F_i(t)$ is given by Eq.(5.8). From this relation, using $F_i^*(s) = 1/(s+\lambda_j)$ by Eq.(5.8), we obtain

$$P_{ij}^*(s) = \frac{\delta_{ij} + h_{ij}^*(s)}{s+\lambda_j} \quad \text{for } i = 1,2,3, \quad j = 1,\cdots,5,$$
$$P_{ij}^*(s) = \delta_{ij} \quad \text{for } i = 4,5, \quad j = 1,\cdots,5.$$

The probability distribution $\{P_{ij}(t)\}$ of states at t is obtained by the Laplace inverse transform of $P_{ij}^*(s)$. This Laplace inverse transform is classified by four cases according to λ_j and a_n. If the initial distribution $\mathbf{p}^0$ is given, the probability distribution of states

$$\mathbf{P}(t) = \{\Sigma_i p_{ij}^0 P_{ij}(t)\}$$

is calculated as follows.

Case 1: λ_j, $j=1,\cdots,5$, and a_n, $n=1,2,3$, are not equal to each other. Let λ_j be a_4.

(5.14a)
$$\mathbf{P}(t) = \sum_{n=1}^{4} \mathbf{N}(-a_n)\exp(-a_n t) + \mathbf{M}(t),$$
$$\mathbf{N}(x) = \frac{B_1 x^2 + B_2 x + B_3}{d(x)},$$
$$d(x) = 4x^3 + 3x^2 \Sigma_n a_n + 2x\Sigma_{n<k} a_n a_k + \Sigma_{n<k<m} a_n a_k a_m,$$
$$\mathbf{M}(t) = \{p_j^0 \exp(-\lambda_j t)\}.$$

Case 2: Two λ_j and a_n are equal (denoted as a_0), and another two terms are not equal (denoted as a_* and a_{**}).

(5.14b)
$$\mathbf{P}(t) = \frac{\mathbf{B}_1 a_*^2 - \mathbf{B}_2 a_* + \mathbf{B}_3}{(a_* - a_0)^2 (a_{**} - a_*)} \exp(-a_* t)$$
$$+ \frac{\mathbf{B}_1 a_{**}^2 - \mathbf{B}_2 a_{**} + \mathbf{B}_3}{(a_{**} - a_0)^2 (a_* - a_{**})} \exp(-a_{**} t)$$
$$+ \left[\frac{\mathbf{C}_1}{(a_* - a_0)^2 (a_{**} - a_0)^2} \right.$$
$$\left. + \frac{\mathbf{B}_1 a_0^2 - \mathbf{B}_2 a_0 + \mathbf{B}_3}{(a_* - a_0)(a_{**} - a_0)} t \right] \exp(-a_0 t) + \mathbf{M}(t),$$
$$\mathbf{C}_1 = a_0^2 \{\mathbf{B}_1 (a_* + a_{**}) - \mathbf{B}_2\} - 2a_0 (\mathbf{B}_1 a_* a_{**} - \mathbf{B}_3)$$
$$+ \mathbf{B}_2 a_* a_{**} - \mathbf{B}_3 (a_* + a_{**}).$$

Case 3: Two pairs of λ_j and a_n are equal. (These value are denoted as a_0 and a_{00}.)

(5.14c)
$$\mathbf{P}(t) = \{\mathbf{C}_2 + \mathbf{C}_3(a_0) t\} \exp(-a_0 t)$$
$$+ \{-\mathbf{C}_2 + \mathbf{C}_3(a_{00}) t\} \exp(-a_{00} t) + \mathbf{M}(t),$$
$$\mathbf{C}_2 = \frac{2\mathbf{B}_1 a_0 a_{00} - \mathbf{B}_2 (a_0 + a_{00}) + 2\mathbf{B}_3}{(a_0 - a_{00})^3},$$
$$\mathbf{C}_3(x) = \frac{\mathbf{B}_1 x^2 - \mathbf{B}_2 x + \mathbf{B}_3}{(a_0 - a_{00})^2}.$$

Case 4: $\lambda_j = a_1 = a_2$ or $\lambda_j = a_1 = a_3$ (denoted as a_0, and the other is denoted as a_*).

(5.14d)
$$\mathbf{P}(t) = \{\mathbf{C}_4 + \frac{\mathbf{B}_1 a_0 (a_0 - 2a_*) + \mathbf{B}_2 a_* - \mathbf{B}_3}{(a_* - a_0)^2} t$$
$$+ \frac{\mathbf{B}_1 a_0^2 - \mathbf{B}_2 a_0 + \mathbf{B}_3}{2(a_* - a_0)^2} t^2 \} \exp(-a_0 t) - \mathbf{C}_4 \exp(-a_* t) + \mathbf{M}(t).$$
$$\mathbf{C}_4 = \frac{\mathbf{B}_1 a_*^2 - \mathbf{B}_2 a_* + \mathbf{B}_3}{(a_* - a_0)^3}.$$

The process is absorbed into State 4 or 5 sooner or later. These probabilities are given by

$$P_4(\infty) = \frac{(p_1^0 + p_2^0) b_1}{b_3},$$
$$P_5(\infty) = p_5^0 + \frac{(p_1^0 + p_2^0) b_2}{b_3}.$$

If the process starts from State 1 ($p_1^0 = 1$) or 2 ($p_2^0 = 1$), these probabilities are

$$P_{14}(\infty) \text{ or } P_{24}(\infty) = \frac{b_1}{b_3},$$
$$P_{15}(\infty) \text{ or } P_{25}(\infty) = \frac{b_2}{b_3}.$$

4. The Expected Time of Absorption

The conditional expected time of absorption $E(T_{ij})$ given that the initial state was i and the process is absorbed into State j (j=4 or 5) is calculated by

$$E(T_{ij}) = - \frac{1}{P_{ij}(\infty)} \left[\frac{dh_{ij}^*(s)}{ds} \right]_{s=0}.$$

Substituting $h_{ij}^*(s)$ and $P_{ij}(\infty)$ into this relation, we obtain

$$E(T_{14}) = \frac{1}{\lambda_1} + \frac{b_1 b_4 - b_3 \lambda_{24}}{b_1 b_3},$$

$$E(T_{15}) = \frac{1}{\lambda_1} + \frac{b_2 b_4 - b_3 \lambda_{25}}{b_2 b_3},$$

$$E(T_{24}) = \frac{b_1 b_4 - b_3 \lambda_{24}}{b_1 b_3},$$

$$E(T_{25}) = \frac{b_2 b_4 - b_3 \lambda_{25}}{b_2 b_3}.$$

The expected time $E(T)$ before absorption into State 4 or 5 given the initial state distribution $\mathbf{p}^0$ is

$$E(T) = \Sigma_i p_i^0 \{\Sigma_{j=4,5} P_{ij}(\infty) E(T_{ij})\}$$

$$= \frac{p_1^0}{\lambda_1} + (p_1^0 + p_2^0) \frac{\lambda_2 p_{23} + \lambda_3 (1 - p_{33})}{b_3}.$$

The variance of the absorbing time T_{ij} is calculated by

$$\sigma^2(T_{ij}) = \frac{1}{P_{ij}(\infty)} \left\{ \frac{d^2 h_{ij}^*(s)}{ds^2} \right\}_{s=0} - E^2(T_{ij}).$$

Next we will consider a situation in which the search time is limited to t_S. In this case, since the target is an appearing and disappearing target, the starting time of the search should be determined optimally. We assume that the detection probability of the target in t_S is adopted as the measure of effectiveness of the search. We define the following quantities:

$$D_1 = \frac{p_1^0 \lambda_A^2}{\lambda_A - \lambda_D} \{\lambda_{24} D_3 + (b_1 + \lambda_D \lambda_{24}) D_4 + \lambda_D b_1 D_5\},$$

$$D_2 = \frac{\lambda_D}{\lambda_A - \lambda_D} \{p_1^0 \lambda_A + (\lambda_A - \lambda_D) p_2^0\}$$

$$\times \{\lambda_{24} D_3 + (b_1 + \lambda_A \lambda_{24}) D_4 + \lambda_A b_1 D_5\},$$

where D_3, D_4, and D_5 are defined as follows.

Case 1: $a_1 \neq a_2$, $a_1 \neq a_3$.

$$D_3 = \sum_{n=1}^{3} \frac{a_n \{1 - \exp(-a_n t_S)\}}{3a_n^2 - 2a_n \Sigma_k a_k + \Sigma_{k<m} a_k a_m},$$

$$D_4 = -\sum_{n=1}^{3} \frac{1 - \exp(-a_n t_S)}{3a_n^2 - 2a_n \Sigma_k a_k + \Sigma_{k<m} a_k a_m},$$

$$D_5 = \sum_{n=1}^{3} \frac{1 - \exp(-a_n t_S)}{a_n \{3a_n^2 - 2a_n \Sigma_k a_k + \Sigma_{k<m} a_k a_m\}}.$$

Case 2: $a_1 = a_2$ or $a_1 = a_3$.

$$D_3 = \frac{a_* \{\exp(-a_0 t_S) - \exp(-a_* t_S)\}}{(a_* - a_0)^2} - \frac{a_0 t_S \exp(-a_0 t_S)}{a_* - a_0},$$

$$D_4 = \frac{\exp(-a_* t_S) - \exp(-a_0 t_S)}{(a_* - a_0)^2} + \frac{t_S \exp(-a_0 t_S)}{a_* - a_0},$$

$$D_5 = \frac{1 - \exp(-a_* t_S)}{a_* (a_* - a_0)^2} + \frac{(a_* - 2a_0)\{1 - \exp(-a_0 t_S)\}}{a_0^2 (a_* - a_0)^2}$$

$$- \frac{t_S \exp(-a_0 t_S)}{a_0 (a_* - a_0)}.$$

The following theorem gives an optimal starting time t_0^* of the search that maximizes the detection probability in $[t_0^*, t_0^*+t_S]$.

Theorem 5.1. *Suppose $\lambda_A \neq \lambda_D$. A necessary condition for the optimal starting time being positive is*

$$D_1 > D_2 > 0, \quad \text{if} \quad \lambda_A > \lambda_D,$$
$$0 > D_1 > D_2, \quad \text{if} \quad \lambda_A < \lambda_D.$$

If these conditions are satisfied, the optimal starting time t_0^* is given by

$$t_0^* = \frac{1}{\lambda_A - \lambda_D} \log \frac{D_1}{D_2}.$$

Proof. Suppose the initial distribution of the target is given by $\mathbf{p}^0$ and the search is conducted in $[t_0, t_0+t_S]$. By setting $b_k(r) = b_k{}^0(r) = 0$, $k=S, T$, in Eqs.(5.14a)~(5.14d), we can calculate the distribution of the target at t_0 as follows.

$$p_1(t_0) = p_1{}^0\exp(-\lambda_A t_0),$$
$$p_2(t_0) = \{p_2{}^0 + \frac{p_1{}^0\lambda_A}{\lambda_A - \lambda_D}\}\exp(-\lambda_D t_0) - \frac{p_1{}^0\lambda_A}{\lambda_A - \lambda_D}\exp(-\lambda_A t_0),$$
$$p_3(t_0) = p_4(t_0) = 0,$$
$$p_5(t_0) = p_5{}^0 + p_2{}^0(1-\exp(-\lambda_D t_0)) + p_1{}^0\{1 - \frac{\{\lambda_A\exp(-\lambda_D t_0) - \lambda_D\exp(-\lambda_A t_0)\}}{\lambda_A - \lambda_D}\}.$$

The detection probability of the target in $[t_0, t_0+t_S]$ is obtained by substituting $p_i(t_0)$ into $p_i{}^0$ of $P_4(t)$ given by Eqs.(5.14a)~(5.14d) and setting $t = t_S$,

$$P_4(t_0+t_S) = p_1(t_0)\{\lambda_1\lambda_{24}D_4 + \lambda_1 b_1 D_5\} + p_2(t_0)\{\lambda_{24}D_3 + (b_1 + \lambda_1\lambda_{24})D_4 + \lambda_1 b_1 D_5\}.$$

Differentiating $P_4(t_0+t_S)$ with respect to t_0, we obtain

$$\frac{dP_4(t_0+t_S)}{dt_0} = D_1\exp(-\lambda_A t_0) - D_2\exp(-\lambda_D t_0).$$

Setting $dP_4(t_0+t_S)/dt_0 = 0$, we obtain the theorem. □

5.3. Forestalling Detection in Barrier Search

In this section, a two-sided search situation between a target in transit along a lane and a searcher patrolling in ambush is considered. It is assumed that the searcher and target are searching for each other continuously and the target can evade the searcher if it forestalls the searcher in detection. Therefore, the searcher must detect the target before the target detects the searcher. The search situation dealt with in this section is as follows.

1. A target is expected to come along a lane at the fixed speed u, but when it comes is completely unknown to the searcher. The width of the lane B is far larger than the range of the detection device, and the searcher wants to ambush the target on a line perpendicular to the lane. The name of barrier is given to the line. The point through which the target crosses the barrier line is also unknown to the searcher; it is assumed to be uniformly distributed on $[0, B]$.

2. The searcher and the target search for each other continuously with detection devices characterized by instantaneous detection rates $k_S b(r)$ and $k_T b(r)$, respectively, where r is the distance between the searcher and the target.

3. The searcher selects one of the four barrier search patterns described:

(1) The fixed-point barrier search: The searcher takes a position at the midpoint of the barrier line and tries to find the target crossing the barrier.

(2) The crossover barrier patrol: The searcher patrols in a bow tie-shaped pattern with speed v in such a manner as the following. The searcher starts at a point on a side of the lane and goes across the lane straight with lead angle $\sin^{-1}(\rho)$, where $\rho = u/v$ is the speed ratio. When he reaches the opposite side of the lane, he changes his course and proceeds up the lane, upsweep distance being d $(= \rho B/\sqrt{1-\rho^2})$. The second half of the period is similar to the first, the searcher goes back across the lane followed by the upsweep. The patrolling continues periodically. Here, we assume $v > u$.

(3) The back-and-forth barrier patrol: The searcher patrols on the barrier line by going forward and backward repeatedly.

(4) The random search: A search area (width $B \times$ depth L) is selected on the lane, and the searcher searches for the target randomly in the area.

We derive the forestalling detection probability for these four barrier patterns.

Let us consider a coordinate system moving with the target. Let Z be the relative course of the searcher in this coordinate system and $r(t|Z)$ be the distance between the searcher and the target at t.

Theorem 5.2. *Let $m_k b(r)$ be the instantaneous detection probability of k and $P_k(t)$ be the probability of k ($k=S$ or T) forestalling j ($j=T$ or S) in detection in search until t. We have*

$$P_k(t) = \frac{m_k}{m_S+m_T}\left[1-\exp\left\{-\int_0^t (m_S+m_T)\, b(r(y|Z))\, dy\right\}\right]. \tag{5.15}$$

Proof. Let $q(t)$ be the nondetection probability in the search until t. Since $q(0) = 1$, we have

$$q(t) = \exp\{-\int_0^t (m_S+m_T)\, b(r(y|Z))\, dy\}.$$

Therefore, the probability density $p_k(t)$ of k forestalling j at t is given by

$$p_k(t) = m_k b(r(t|Z)) \exp\{-\int_0^t (m_S+m_T)\, b(r(y|Z))\, dy\}.$$

Integrating $p_k(x)$ in $x \in [0, t]$ on Z, we obtain Eq.(5.15). □

The detection probability $P_0(t|m_S)$ of a one-sided search with detection function, $m_S b(r)$, is easily obtained by setting $m_T = 0$ in Eq.(5.15) as

$$P_0(t|m_S) = 1 - \exp\{-\int_0^t m_S b(r(y|Z))\, dy\}.$$

We can easily find the following relation between $P_0(t|m_s)$ and $P_k(t)$ from this formula and Theorem 5.2.

Corollary 5.3. *The forestalling detection probability $P_k(t)$ is calculated from the one-sided model by*

$$P_k(t) = \frac{m_k}{m_S+m_T} P_0(t|m_S+m_T). \tag{5.16}$$

It should be noted that Eq.(5.16) is valid for any relative motion Z if the detection functions of both sides have similar forms with different constants, $m_k b(r)$, $k = S, T$. Applying Corollary 5.3, we can obtain the forestalling detection probability for both sides in the two-sided search from the one-sided search model.

In the following, assuming the inverse cube detection law of the detection devices, $b(r)=1/r^3$, we shall derive the forestalling detection probability for the four patterns of barrier search mentioned before.

1. The Fixed-Point Barrier Search

Considering an infinite straight course Z with lateral range x from the searcher in Eq(5.15), we obtain the lateral range curve $PL_k(x)$ in the two-sided search:

$$PL_k(x) = \frac{m_k}{m_S+m_T} [1-\exp\{-\frac{2(m_S+m_T)}{ux^2}\}].$$

Since the point where the target crosses the barrier line is assumed to be uniformly distributed, the forestalling detection probability PP_k by k in the fixed barrier search is given by

$$PP_k = \frac{2}{B}\int_0^{B/2} PL_k(x)\, dx$$
$$= \frac{m_k}{m_S+m_T} [1- \frac{2}{B}\int_0^{B/2} \exp\{-\frac{2(m_S+m_T)}{ux^2}\} dx].$$

Changing the variable to $y = 2\sqrt{(m_S+m_T)/u}\,/x$ and integrating by parts, the integration results in an integration of the normal density function. We apply the formula of Williams' approximation for the normal integral,

$$\frac{1}{\sqrt{2\pi}}\int_{-c}^{c}\exp\left(-\frac{y^2}{2}\right)dy \fallingdotseq \left\{1-\exp\left(-\frac{2c^2}{\pi}\right)\right\}^{1/2}.$$

Then we have

$$PP_k = \frac{m_k}{m_S+m_T}\left[1-\exp\left\{-\frac{8(m_S+m_T)}{B^2u}\right\}\right.$$

$$\left.+\frac{2}{B}\left\{\frac{2\pi\sqrt{m_S+m_T}}{u}\right\}\left\{1-\left\{1-\exp\left(-\frac{32(m_S+m_T)}{\pi B^2 u}\right)\right\}^{1/2}\right\}\right].$$

2. The Crossover Barrier Patrol

Koopman (1946) gave an approximation of the detection probability $PC_0(m_s)$ of the crossover barrier patrol in a one-sided search.

$$PC_0(m_S) = \left(\frac{2}{\pi}\right)^{1/2}\int_0^a \exp\left(-\frac{y^2}{2}\right)dy,$$

$$a = \frac{2\pi}{D}\left(\frac{m_S}{w}\right)^{1/2},\qquad \rho = \frac{u}{v}\ (<1),\qquad w^2 = v^2-u^2,$$

$$D = (1+\rho)d,\qquad d = \frac{B\rho}{\sqrt{1-\rho^2}}.$$

From Corollary 5.3, we obtain the forestalling detection probability PC_k by k, $k = S, T$,

$$PC_k = \frac{m_k}{m_S+m_T}\left(\frac{2}{\pi}\right)^{1/2}\int_0^{a'}\exp\left(-\frac{y^2}{2}\right)dy,$$

$$a' = \frac{2\pi}{B(\rho+1)}\left(\frac{m_S+m_T}{\rho u}\right)^{1/2}(1-\rho^2)^{1/4}.$$

Applying Williams' formula, we have

$$PC_k = \frac{m_k}{m_S+m_T}\left[1-\exp\left\{-\frac{8\pi\sqrt{1-\rho^2}\,(m_S+m_T)}{u\rho(\rho+1)^2B^2}\right\}\right]^{1/2}.$$

3. The Back-and-Forth Barrier Patrol

Applying similar approximation to Koopman's model for the crossover barrier patrol, Iida (1988b) obtained the detection probability for the back-and-forth barrier patrol in a one-sided search.

$$PB_0(m_S) = 1-\left[1-\left(\frac{2}{\pi}\right)^{1/2}\int_0^{a''}\exp\left(-\frac{y^2}{2}\right)dy\right]^2,$$

$$a'' = \frac{\pi}{B}\left(\frac{m_S}{u\rho}\right)^{1/2}(\rho^2+1)^{1/4}.$$

Therefore, we can derive the forestalling detection probability PB_k by k from Corollary 5.3 and Williams' formula as follows:

$$PB_k = \frac{m_k}{m_S+m_T}\left[1-\left\{1-\left[1-\exp\left(-\frac{2\pi\sqrt{1+\rho^2}\,(m_S+m_T)}{u\rho B^2}\right)\right]^{1/2}\right\}^2\right].$$

4. The Random Search

In this case, the searcher sets a search area $B \times L$ on the lane and searches for the target with the random search in the area. From the random search formula, we have

(5.17) $$PR_0(t|m_S) = 1 - \exp(-\frac{Qt}{BL}),$$

where Q is the sweep rate of the searcher. Considering an encounter with a track angle θ between the searcher and the target, we have the conditional sweep rate as

$$Q(\theta) = 2\sqrt{2\pi m_S w(\theta)},$$
$$w(\theta)^2 = v^2+u^2-2uv\cos\theta.$$

Since the track angle θ is uniformly distributed on $[0,2\pi]$, the sweep rate Q is

$$Q = E[Q(\theta)] = 2\sqrt{2\pi m_S v}\ f(\rho),$$

where

(5.18) $$f(\rho) = \frac{1}{2\pi}\int_0^{2\pi}\{\rho^2+1-2\rho\cos\theta\}^{1/4}d\theta.$$

It takes time $t = L/u$ for the target to go through the search area. Substituting Q and t into Eq.(5.17), we have the detection probability of the target in the one-sided search.

$$PR_0(m_S) = 1 - \exp(-\frac{2\sqrt{2\pi m_S v}\ f(\rho)}{B\ u}).$$

Therefore, the forestalling detection probability PR_k by k is obtained from Corollary 5.3,

$$PR_k = \frac{m_k}{m_S+m_T}[1-\exp(-\frac{2\sqrt{2\pi(m_S+m_T)v}\ f(\rho)}{B\ u})].$$

Remark 5.1. The function $f(\rho)$ given by Eq.(5.18) is called the dynamic enhancement of the sweep rate. The value of this function means the enhancement of the sweep rate caused by the random movement of the target. This function is strictly increasing with respect to the speed ratio ρ $(= u/v)$ and $f(\rho) > 1$ for $\rho > 0$. Of course, this function depends on the detection function. $f(\rho)$ for the detection function of cookie-cutter type was derived by Koopman (1946) as follows:

(5.19) $$f(\rho) = \frac{1}{2\pi}\int_0^{2\pi}\{\rho^2+1-2\rho\cos\theta\}^{1/2}d\theta.$$

The function is essentially the complete elliptic integral of the second kind. It is interesting to note that Eqs.(5.18) and (5.19) are very similar. Fig. 5.1 shows the curves $f(\rho)$ given by Eqs.(5.18) and (5.19).

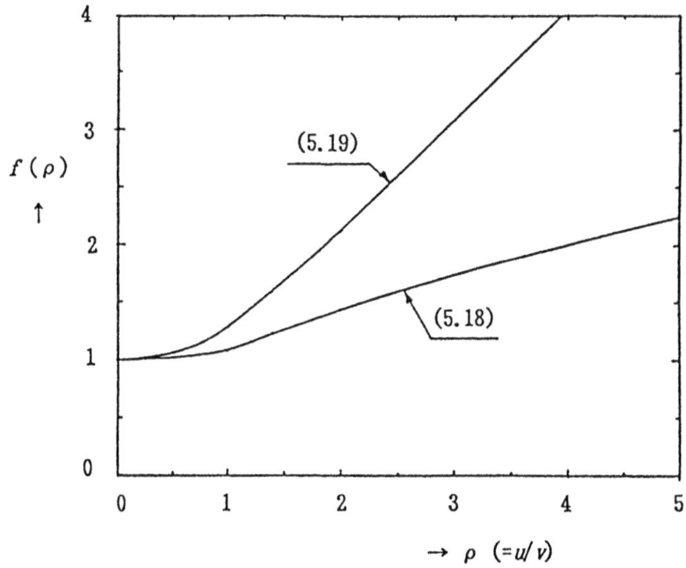

Figure 5.2. Dynamic enhancement.

Remark 5.2. To illustrate the forestalling detection probability for each pattern, PP_S, PC_S, PB_S and PR_S are calculated for $m_S/m_T = 1$, $\rho = 0.4$. Fig. 5.3 shows these curves, where the abscissa is the ratio κ of the barrier width to the sweep width of the searcher:

$$\kappa = \frac{B}{W_S} = \frac{B}{2\sqrt{2\pi m_S/v}} .$$

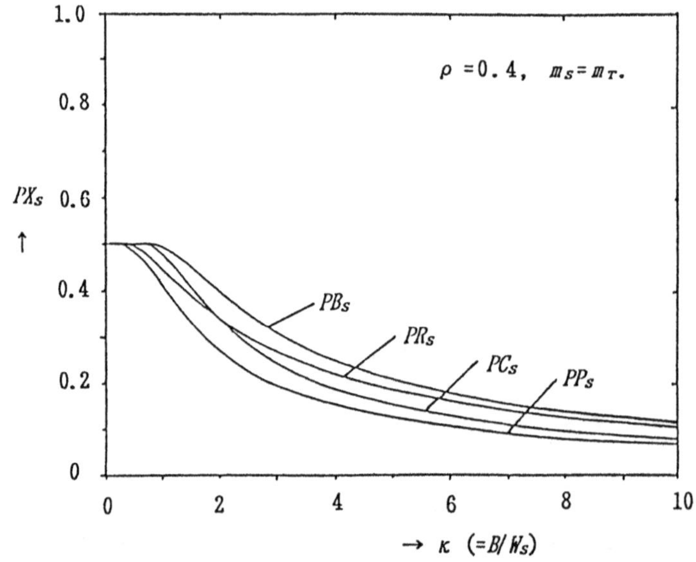

Figure 5.3. Forestalling probability.

As seen in Fig. 5.3, if the searcher wishes to maximize the forestalling detection probability, the back-and-forth barrier patrol is most efficient. However, it should be noted that this conclusion is valid only for the relatively high speed target, ρ - 0.4.

Remark 5.4. In this section, we obtain the forestalling detection probability in barrier search situation. Using these formula, Iida and Nakamura (1989) derived the overall probability of detecting the target by the searcher in barrier patrol. In their model, the searcher and the target are searching for each other by detection devices characterized by the inverse cube law of detection. When the target forestalls the searcher in detection before the barrier line, the target is assumed to penetrate into the barrier line selecting the best timing and the best point. The conditional probability of detecting the target is derived by a game theoretical analysis. On the other hand, if the target forestalls the searcher after the barrier is passed, it continues its transit quietly. In this case, the detectability of the target by the searcher decreases and the searcher's sensor parameter m_S of the inverse cube law is assumed to be reduced to βm_S, $0 < \beta \leq 1$. Taking these effects of the forestalling detection by the target into account, they derive the overall detection probabilities for the four patterns of barrier patrol; fixed-point barrier search, crossover barrier patrol, back-and-forth barrier patrol and random search. Furthermore, applying these models, Iida and Nakamura (1990) investigated the optimal pattern of barrier patrol maximizing the overall detection probability. Evaluating the overall detection probability of each pattern for a set of system parameters; W_T/W_S, β, ρ $(=u/v)$ and κ $(=B/W_S)$, the optimal pattern is sought and several charts that indicate the optimal pattern in the space of system parameters are obtained.

Bibliography

Ahlswede R., and Wegener, I. (1987), *Search problem,* John Wiley & Sons Ltd., New York.

Baston, V.J., and Bostock, F.A. (1987). "A continuous game of ambush," *Naval Research Logistics, 34,* 645-654.

Black, W.L. (1965). "Discrete sequential search," *Information and Control, 8,* 159-162.

Brown, S.S. (1980). "Optimal search for a moving target in discrete time and space," *Operations Research, 28,* 1275-1289.

Charnes, A., and Cooper, W.W. (1958). "The theory of search: Optimum distribution of search effort," *Management Science, 5,* 44-50.

Chew, M.C.Jr. (1967). "A sequential search procedure," *The Annals of Mathematical Statistics, 38,* 494-502.

Chew, M.C.Jr. (1973). "Optimal stopping in a discrete search problem," *Operations Research, 21,* 741-747.

Chudnovsky, D.V. and Chudnovsky, G.V. (Eds.), (1989), *Search theory,* Marcel Dekker Inc., New York.

Danskin, J.M. (1968). "A helicopter versus submarine search game," *Operations Research, 16,* 509-517.

de Guenin, J. (1961). "Optimum distribution of effort: An extension of the Koopman basic theory," *Operations Research, 9,* 1-7.

Dobbie, J.M. (1963). "Search theory: A sequential approach," *Naval Research Logistics Quarterly, 10,* 323-334.

Dobbie, J.M. (1968). "A survey of search theory," *Operations Research, 16,* 525-537.

Dobbie, J.M. (1973). "Some search problems with false contacts," *Operations Research, 21,* 907-925.

Dobbie, J.M. (1974). "A two-cell model of search for a moving target," *Operations Research, 22,* 79-92.

Eagle, J.N. (1984). "The optimal search for a moving target when the search path is constrained," *Operations Research, 32,* 1107-1115.

Engel, J.H. (1957). "Use of clustering in mineralogical and other surveys," *Proceedings of First International Conference on Operations Research,* 176-192.

Enslow, P.H.Jr. (1966). "A bibliography of search theory and reconnaissance theory literature," *Naval Research Logistics Quarterly, 13,* 177-202.

Gal, S. (1980). *Search Games,* Academic Press, New York.

Giammo, T.P. (1963). "On a probability of success in a sudden death search with intermittent moves confined to a finite area," *SIAM. Review, 5,* 41-51.

Gilbert, E.N. (1959). "Optimal search strategies," *SIAM. J. Appl. Math. 7,* 413-424.

Gittins, J.C., and Roberts, D.M. (1979). "The search for an intelligent evader concealed in one of an arbitrary number of regions," *Naval Research Logistics Quarterly, 26,* 651-666.

Gluss, B. (1961). "Approximately optimal one-dimensional search policies in which search costs vary through time," *Naval Research Logistics Quarterly, 8*, 277-283.
Haley, K.B., and Stone, L.D. (Eds.), (1980). *Search Theory and Applications,* Plenum Press, New York.
Hellman, O. (1970). "On the effect of a search upon the probability distribution of a target whose motion is a diffusion process," *The Annals of Mathematical Statistics, 41*, 1717-1724.
Hellman, O. (1971). "Optimal search for a randomly moving object in a special case," *Journal of Applied Probability, 8*, 606-611.
Hellman, O. (1972). "On the optimal search for a randomly moving target," *SIAM. J. Appl. Math. 22*, 545-552.
Iida, K. (1972). "The optimal distribution of searching effort for a moving target," (in Japanese), *Keiei Kagaku, 16*, 204-215.
Iida, K. (1974). "Current topics in OR: Search theory," (in Japanese), *Keiei Kagaku, 18*, 239-243.
Iida, K. (1982). "Optimal stopping rule in a two-box continuous search," *Memoirs of the National Defense Academy, 22*, 271-284.
Iida, K. (1984). "Optimal search and stop in continuous search process," *Journal of the Operations Research Society of Japan, 27*, 1-30.
Iida, K. (1985). "A two-sided search model considering detection forestalled by the target. Ⅲ: A case of searching for an appearing / disappearing target," (in Japanese), *Scientific and Engineering Reports of the National Defense Academy, 23*, 83-105.
Iida, K. (1988a). "Optimal whereabouts search for a target with a random lifetime," *Memoirs of the National Defense Academy, 28*, 71-85.
Iida, K. (1988b). "The forestalling detection probability of the target in two-sided barrier patrol," (in Japanese), *Scientific and Engineering Reports of the National Defense Academy, 26*, 43-61.
Iida, K. (1989a). "Optimal stopping of a contact investigation in two-stage search," *Mathematica Japonica, 34*, 169-190 .
Iida, K. (1989b). "Optimal search plan minimizing the expected risk of the search for a target with conditionally deterministic motion," *Naval Research Logistics, 36*, 597-613.
Iida, K., and Hiramoto, S. (1984). "A suboptimal distribution of searching effort for a moving target," (in Japanese), *Scientific and Engineering Reports of the National Defense Academy, 22*, 169-196.
Iida, K., and Hozaki R. (1988). "The optimal search plan for a moving target minimizing the expected risk," *Journal of the Operations Research Society of Japan, 31*, 294-320.
Iida, K., and Kanbashi, T. (1983). "The optimal whereabouts search policy minimizing the expected risk," *Memoirs of the National Defense Academy, 23*, 187-209.
Iida, K., and Nakamura, T. (1989). "Mathematical models of two-sided barrier search assuming possible forestalling detection by the target," (in Japanese), *Scientific and Engineering Reports of the National Defense Academy, 27*, 215-230.
Iida, K., and Nakamura, T. (1990). "Optimal pattern of barrier patrol in two-sided barrier search," *Defense Technology Journal, 10*, 2, 35-50.
Johnson, S.M. (1964). "A search game," RM-3717-PR (Rand Corp. Report).
Kadane, J.B. (1969). "Discrete search and the Neyman Peason lemma," *J. Math. Analysis and Appl. 26*, 609-623.
Kadane, J.B. (1971). "Optimal whereabouts search," *Operations Research, 19*, 894-904.
Kan, Y.C. (1974). "A counter example for an optimal search and stop model," *Operations Research, 22*, 889-892.
Kan, Y.C. (1977). "Optimal search for a moving target," *Operations Research, 25*, 864-870.

Kisi, T. (1966). "On an optimal searching schedule," *Journal of the Operations Research Society of Japan, 8,* 53-65.

Kisi, T. (1979). "Optimal stopping of the investigating search," in *Search Theory and Applications,* 255-260, Plenum Press, New York.

Kisi, T., and Iida, K. (1971). "Present state of the theory of search," (in Japanese), *Keiei Kagaku, 15,* 13-28.

Kisi, T., and Tatsuno, K. (1981). "Adaptive search and stop with uncertain sensor capability," *Journal of the Operations Research Society of Japan, 24,* 189-201.

Koopman, B.O. (1946). "*Search and screening,*" OEG Rep. No.56, Center for Naval Analysis, Rosslyn, Va.

Koopman, B.O. (1956a). "The theory of search I. Kinematic bases," *Operations Research, 4,* 324-346.

Koopman, B.O. (1956b). "The theory of search II. Target detection," *Operations Research, 4,* 503-531.

Koopman, B.O. (1957). "The theory of search III. The optimum distribution of searching effort," *Operations Research, 5,* 613-626.

Koopman, B.O. (1980). *Search and Screening* (2nd ed.), Pergamon Press, New York.

Lössner, U., and Wegener, I. (1982). "Discrete sequential search with positive switch cost," *Math. of Opns. Res., 7,* 426-440.

Matula, D. (1964). "A periodic optimal search," *American Mathematics Monthly, 71,* 15-21.

Mela, D.F. (1961). "Information theory and search theory as special case of decision theory," *Operations Research, 9,* 907-909.

Moore, M.L. (1970) "A review of search and reconnaissance theory literature," Systems Research Laboratory (The Univ. of Michigan), SRL-2147-TR-70-1 (AD-700333).

Nakai, T. (1973). "A model of search for a target moving among three boxes: Some special cases," *Journal of the Operations Research Society of Japan, 6,* 151-162.

Nakai, T. (1980). "A search model for a moving target in which the decision wait is permitted," *Mathematica Japonica,* 25, 597-605.

Nakai,T. (1982). "Optimal search for an object with a random lifetime," *Journal of the Operations Research Society of Japan, 25,* 175-192.

Nakai, T. (1986a). "A sequential evasion-search game with a goal," *Journal of the Operations Research Society of Japan, 29,* 113-121.

Nakai, T. (1986b). "A search game with one object and two searchers," *Journal of Applied Probability, 22,* 696-707.

Nakai, T. (1987). "Continuous-time search model with improvements of detection rates," *Journal of the Operations Research Society of Japan, 30,* 316-333.

Neuts, M.F. (1963). "A multistage search game," *SIAM.J.Appl.Math. 11,* 502-507.

Norris, R.C. (1962). "Studies in search for a conscious evader," Rep. No.279, Lincoln Laboratory M.I.T., (AD-294832).

Onaga, K. (1971). "Optimal search for detecting a hidden object," *SIAM. J.Appl.Math. 20,* 298-318.

Pollock, S.M. (1970). "A simple model of search for a moving target," *Operations Research, 18,* 883-903.

Purshiheimo, U. (1977). "On the optimal search for a target whose motion is conditionally deterministic with stochastic initial condition on location and parameters," *SIAM.J.Appl.Math. 32,* 105-114.

Richardson, H.R. (1987). "Search theory," Center for Naval Analyses, N00-014-83-C-0725 (AD-A177493).

Richardson, H.R., and Belkin, B. (1972). "Optimal search with uncertain sweep width," *Operations Research, 20,* 764-784.

Ross, S.M. (1969). "A problem in optimal search and stop," *Operations Research, 17,* 984-992.

Ruckle, W.H. (1981). "Ambushing random walks. II: Continuous models," *Operations Research, 29*, 108-120.
Ruckle, W.H., Fennel, R., Holms, P.T., and Fennemore, C. (1976). "Ambushing random walks. I: Finite models," *Operations Research, 24*, 314-324.
Ruckle, W.H., and Reay, J.R. (1981). "Ambushing random walks. III: More continuous models," *Operations Research, 29*, 121-129.
Sakaguchi, M. (1973). "Two-sided search game," *Journal of the Operations Research Society of Japan, 16*, 207-225.
Saretsalo, L. (1973). "On the optimal search for a target whose motion is a Markov process," *Journal of Applied Probability, 10*, 847-856.
Smith, M.W., and Walsh, J.E. (1971). "Optimum sequential search with discrete locations and random acceptance errors," *Naval Research Logistics Quarterly, 19*, 159-167.
Staroverov, O.V. (1963). "On a searching problem," *Theory of Probability and Its Applications, 8*, 184-187.
Stone, L.D. (1975). *Theory of Optimal Search*, Academic Press, New York.
Stone, L.D. (1977). "Search for targets with generalized conditionally deterministic motion," *SIAM. J. Appl. Math.* 33, 456-468.
Stone, L.D. (1979). "Necessary and sufficient conditions for optimal search plans for moving targets," *Math. of Opns. Res.*, *4*, 431-440.
Stone, L.D. (1983). "The process of search planning: Current approaches and continuing problems," *Operations Research, 31*, 207-233.
Stone, L.D., and Kadane, J.B. (1981). "Optimal whereabouts search for a moving target," *Operations Research, 29*. 1154-1166.
Stone, L.D., and Richardson, H.R. (1974). "Search for targets with conditionally deterministic motion," *SIAM. J. Appl. Math. 27*, 239-255.
Stone, L.D., and Stanshine, J.A. (1971). "Optimal search using uninterrupted contact investigation," *SIAM. J. Appl. Math. 20*, 241-263.
Stone, L.D., Stanshine, J.A., and Persinger, C.A. (1972). "Optimal search in the presence of Poisson-distributed false target," *SIAM. J. Appl. Math. 23*, 6-27.
Stromquist, W.R., and Stone, L.D. (1981). "Constrained optimization of functionals with search theory applications," *Math. of Opns. Res., 16*, 518-529.
Tada, K. (1964). "A study of the optimal distribution of searching effort," (in Japanese), *Keiei Kagaku, 7*, 81-86.
Tada, K. (1965). "A study of the optimal distribution of searching effort:(II)," (in Japanese), *Keiei Kagaku, 8*, 218-224.
Tada, K. (1973). *Search Theory*, (in Japanese), Nikkagiren Shuppan, Tokyo.
Tatsuno, K., and Kisi, T. (1982). "ILA stopping rule for a search with uncertain sensor capability," *Journal of the Operations Research Society of Japan, 25*, 363-375.
Tierney, L., and Kadane, J.B. (1983). "Surveillance search for a moving target," *Operations Research, 31*, 720-738.
Tognetti, K.P. (1968). "An optimal strategy for a whereabouts search," *Operations Research, 16*, 209-211.
Washburn, A.R. (1974). "Hide and seek from a fixed base," Rep. No. 55-Ws-79010, Naval Postgraduate School.
Washburn, A.R. (1980a). "On a search for a moving target," *Naval Research Logistics Quarterly, 27*, 315-322.
Washburn, A.R. (1980b). "Search-evasion game in a fixed region," *Operations Research, 28*, 1290-1298.
Washburn, A.R. (1981a). *Search and Detection*, Military Applications Section of Operations Research Society of America, Ketron, Inc., Arlington, Va.
Washburn, A.R. (1981b). "An upper bound useful in optimizing search for a moving target," *Operations Research, 29*, 1227-1230.
Washburn, A.R. (1983). "Search for a moving target: The FAB algorithm," *Operations Research, 31*, 739-750.

Author Index

Numbers in italics refer to the pages on which the complete references are listed.

A

Ahlswede, R., 2, *122.*

B

Baston, V.J., 104, *122.*
Belkin, B., 9, *124.*
Black, W.L., 8, *122.*
Bostock, F.A., 104, *122.*
Brown, S.S., 29, 39, *122.*

C

Charnes, A., 8, *122.*
Chew, M.C.Jr., 9, *122.*
Chudnovsky, D.V., 2, *122.*
Chudnovsky, G.V., 2, *122.*
Cooper, W.W., 8, *122.*

D

Danskin, J.M., 58, 103, *122.*
de Guenin, J., 8, *122.*
Dobbie, J.M., 2, 8, 29, 85, *122.*

E

Eagle, J.N., 30, *122.*
Engel, J.H., 84, *122.*
Enslow, P.H.Jr., 2, *122.*

F

Fennel, R., 104, *124.*
Fennemore, C., 104, *124.*

G

Gal, S., 2, *122.*
Giammo, T.P., 103, *122.*
Gilbert, E.N., 8, *122.*
Gittins, J.C., 103, *122.*
Gluss, B., 8, *123.*

H

Haley, K.B., 2, *123.*
Hellman, O., 29, *123.*
Hiramoto, S., 59, *123.*
Holms, P.T., 104, *124.*
Hozaki R., 30, 33, *123.*

I

Iida, K., 2, 9, 29, 30, 33, 46, 59, 61, 70, 79, 85, 104, 118, 121, *123.*

J

Johnson, S.M., 103, *123.*

K

Kadane, J.B., 8, 30, 61, 68, *123.*
Kan, Y.C., 9, 29, *123.*
Kanbashi, T., 61, 79, *123.*
Kisi, T., 2, 8, 9, 85, *124.*
Koopman, B.O., 1, 2, 8, 103, 118, 119, *124.*

L

Lössner, U., 8, *124.*

M

Matula, D., 8, *124.*
Mela, D.F., 60, *124.*
Moore, M.L., 2, *124.*

N

Nakai, T., 9,29,79,103, *124.*
Nakamura, T. 121, *123.*
Neuts, M.F., 103, *124.*
Norris, R.C., 103, *124.*

O

Onaga, K., 8, *124.*

P

Persinger, C.A., 84, *125.*
Pollock, S.M., 29, *124.*
Purshiheimo, U., 29, *124.*

R

Reay, J.R., 104, *124.*
Richardson, H.R., 2,9,29,48, *124.*
Roberts, D.M., 103, *122.*
Ross, S.M., 9, *124.*
Ruckle, W.H., 104, *125.*

S

Sakaguchi, M., 103, *125.*
Saretsalo, L., 29, *125.*
Smith, M.W., 84, *125.*
Stanshine, J.A., 84, *125.*
Staroverov, O.V., 8, *125.*
Stone, L.D., 1,2,29,30,35,48, 61,84, *125.*
Stromquist, W.R., 30,35, *125.*

T

Tada, K., 2,17, *125.*
Tatsuno, K., 9, *125.*
Tierney, L., 30, *125.*
Tognetti, K.P., 60, *125.*

W

Walsh, J.E., 84, *125.*
Washburn, A.R., 2,30,39,79, 97,103,104, *125.*
Wegener, I., 2,8, *122, 124.*

Subject Index

A

Absorbing state, 105.
Absorption, 113, 114.
Additive, 50, 51.
Alertness, 104, 105.
Algorithm, 30, 39, 43, 61, 79, 93, 98.
Appeared target, 2, 10, 31, 33, 61, 80.
Appearing target, 2, 105.
Application of search theory, 1.
Attributes of search model, 2-4.

B

Balanced search, 17, 21, 23.
Barrier search, 104, 115-121.
  back-and-forth, 116, 118, 121.
  crossover, 116, 118.
  fixed-point, 116, 117.
  random, 116, 119.
Behavior of $\lambda(T)$, 21.
Broad search, 3, 84-87, 104-115.

C

Complete search, 43.
Concave, 32, 36, 37.
Conditionally deterministic motion (C.D.M.) target, 45-46.
Conditionally optimal search plan, 11, 36, 51, 63, 72.
Conditionally optimal stopping time, 22, 67, 75, 76.
Constraint, 11, 31, 34, 62, 71. *See* also restriction.
Contact, 3, 52, 85, 86, 97.
Continuous divisibility of searching effort, 3, 10, 14, 31, 33, 62, 79.
Convexity of
  the expected risk, 13, 38, 64, 72.
  $\lambda(t)$, 14, 66, 73.
Correct guess, 62, 71, 81, 83.
Critical contact surface, 94, 95, 98, 102.
Cumulative searching cost, 34, 47.
Cumulative searching effort, 10, 33, 47, 62, 65.

D

Datum point, 59, 85, 87, 97.
Datum point error, 85, 86, 87, 94.
Datum search, 52, 97.
Detectability, 33, 104, 105, 121.
Detection function
  conditional, 3, 10, 31, 33, 47, 87, 117.
  cookie-cutter type, 58, 119.
  exponential, 3, 8, 10, 14, 43, 52, 62, 79, 101, 103.
  inverse cube law, 103, 109, 117, 121.
  random search, *see* exponential detection function.
  regular, 3, 14, 31, 33, 47, 49, 79.
Detection probability, 3, 18, 31, 34, 47, 115, 117.
Diffusive F.C.D.M. target, 47, 49, 51, 58, 59.
Disappearing target, 2, 105.
Discrete searching effort, 2, 3, 9.
Dynamic enhancement for
  cookie-cutter type, 119.
  inverse cube law, 119.

E

Encounter, 106, 107, 108, 119.
Evasion-and-search problem, 103.
Expected
  lost price versus cost ratio, 77.
  marginal return, 39, 42, 65, 75, 78.
  marginal return versus cost ratio, 13, 21, 39, 65, 68, 75, 77.
  number of transitions, 111-112.
  reward, 21, 74.
  risk, 3, 9, 10, 13, 22, 23, 34, 39, 40, 45, 48, 62, 65, 71.
  searching cost until detection, 3, 21.

Expected (continued)
time of absorption,113-114.
time of detection,3,88.

F

FAB algorithm,30,39,43,79.
Factorable,45,46,47.
False contact,3,84,85,87.
F.C.D.M. target,46-47,49.
Feasible search plan,11,34,88.
Forestalling detection in
barrier search,115-121.
broad search,104-115.
Fractional programming problem,88, 89,90.

G

Gateaux differential,35,36,37,64, 72.

H

Hide-and-search problem,58,103.

I

Immortal target,61,79,80,81.
Influence of the lifetime,80.
Inhomogeneity of the reward,18,58.
Initial state (distribution),110, 112,114,115.
Instantaneous detection rate,104, 116.
Inverse cube law:*see* detection function.
Inverse function,91,93,102.
Investigating search,3,84,86,87.
Investigating search plan,87.

J

Jacobian,45,46.

K

Kernel of Gateaux differential,35, 36,38,64,72.
Kuhn-Tucker theorem,32.

L

Laplace inverse transform,110,112.
Laplace transform,109,110.
Lateral range,107,108,117.
Lifetime of the target,70,81.
Limit time of the search,34,35,42, 44.

M

Marginal detection probability,32, 68,89,92.
Markov chain,43,105,109.
Mean detection rate,92.
Meaning of condition for
optimal distribution of searching effort,13,32, 38,65,74.
optimal guess box,70,78.
optimal investigating search plan,92.
optimal stopping time,21,42,68, 77.
Measure of effectiveness,3,4,9,10, 34,47,60,62,87,114.
Miss one strategy,60,61,68.
Mortal target,70,81,83.
Motion parameter,45,47,53,59.
Moving target,2,29-59.
Myopic search plan,17,32.

N

Necessary and sufficient condition,
optimal investigating search plan,90,91.
optimal whereabouts search plan, 63,72.
R_T-optimal search plan
for a moving target,36,49.
for a stationary target,11.
uniformly P-optimal plan for a F.C.D.M. target,49.
Necessary condition for
optimal box to be guessed,69,70, 78.
optimal starting time of search, 115.
optimal stopping time,18,19,41, 51,67,75,76.
R_T-optimal search plan,36,40,49.
Noisy search,3,84.
Nondetection probability,18,67,75, 116.
Nonlinear programming problem,31.

O

Optimal box to be guessed,69,70,78.

Optimal investigating search plan, 90,91,92.
Optimal pattern of barrier patrol, 121.
Optimal plan region (OPR),24,26, 27,28.
Optimal search plan for
investigating search,84-102.
moving target,29-59.
(F.C.D.M. target,45-59.)
stationary target,8-28.
(two-box problem,22-28.)
whereabouts search for
immortal target,61-70.
mortal target,70-79.
Optimal starting time,16,66,115.
Optimal stopping time,16,18,19,35, 41,51,66,67,68.
Overall detection probability,121.

P

Partial search,43,44.
Path,31,33.
P-optimal search,46,58.
Posterior probability distribution of the target,10,17,18,21,22, 23,76,77.
Properties of
box to be searched,20,68.
conditionally R_T-optimal search plan,15.
critical contact surface,94
detection function of investigation,87.
optimal investigating search plan,94-96.
$\lambda(t)$,14,21,66,73.

Q

Quality of contact,86,95,96,100.
Quantized effort,2,14,79.

R

Random lifetime,9,70.
Random search,3,10,62,105,116,119.
Renewal density function,110-111.
Restriction,48,63,88.See also constraint.
Reward,10,34,39,47,62,70,71.
R-optimal,46,53,55,56,57.
R_T-optimal,35,40,43,51.

S

Saved risk,13,38,65,68,75.
Search-and-rescue,1,60.
Searching effort,2.
Search region,41,55.
Second-stage search,84.
Speed distribution,
Gamma,56.
Rayleigh,56.
triangular,55,59.
uniform,54-55.
Starting time,16,23,66,115.
State distribution,112,113.
State of the search process,105,106.
Stationary target,2,10,61,70.
Stopping time,16,18,22,35,40,42,63, 66,87.
Structure of OPR,24,26,27.
Sweep rate,101,119.
Sweep width,97,103,109.

T

Time lag,52,85,86,87,94,98,102.
Transformation,45,46,59.
Transition probability density function,109.
Transition probability matrix,43.
Transition rate,106,108.
Two-sided search,2,103-121.
barrier search,115-121.
broad search,104-115.
Two-stage search,84.

U

Uniformly P-optimal,48,49,51,58.
Uniformly R-optimal,51.
Unit cost,10,34,47.

V

Variance of the absorbing time,114.
Variational problem,11,35,48,63,71.

W

Whereabouts probability,3,60,61.
Whereabouts search,60,62.
Williams' formula,118.

GPSR Compliance

The European Union's (EU) General Product Safety Regulation (GPSR) is a set of rules that requires consumer products to be safe and our obligations to ensure this.

If you have any concerns about our products, you can contact us on ProductSafety@springernature.com

In case Publisher is established outside the EU, the EU authorized representative is:

Springer Nature Customer Service Center GmbH
Europaplatz 3
69115 Heidelberg, Germany

Batch number: 10704763

Printed by Printforce, the Netherlands